Selected Titles in This Series

706 **Jonathan Brundan, Richard Dipper, and Alexander Kleshchev,** Quantum linear groups and representations of $GL_n(\mathbb{F}_q)$, 2001

705 **I. Moerdijk and J. J. C. Vermeulen,** Proper maps of toposes, 2000

704 **Jeff Hooper, Victor Snaith, and Min van Tran,** The second Chinburg conjecture for quaternion fields, 2000

703 **Erik Guentner, Nigel Higson, and Jody Trout,** Equivariant E-theory for C^*-algebras, 2000

702 **Ilijas Farah,** Analytic guotients: Theory of liftings for quotients over analytic ideals on the integers, 2000

701 **Paul Selick and Jie Wu,** On natural coalgebra decompositions of tensor algebras and loop suspensions, 2000

700 **Vicente Cortés,** A new construction of homogeneous quaternionic manifolds and related geometric structures, 2000

699 **Alexander Fel′shtyn,** Dynamical zeta functions, Nielsen theory and Reidemeister torsion, 2000

698 **Andrew R. Kustin,** Complexes associated to two vectors and a rectangular matrix, 2000

697 **Deguang Han and David R. Larson,** Frames, bases and group representations, 2000

696 **Donald J. Estep, Mats G. Larson, and Roy D. Williams,** Estimating the error of numerical solutions of systems of reaction-diffusion equations, 2000

695 **Vitaly Bergelson and Randall McCutcheon,** An ergodic IP polynomial Szemerédi theorem, 2000

694 **Alberto Bressan, Graziano Crasta, and Benedetto Piccoli,** Well-posedness of the Cauchy problem for $n \times n$ systems of conservation laws, 2000

693 **Doug Pickrell,** Invariant measures for unitary groups associated to Kac-Moody Lie algebras, 2000

692 **Mara D. Neusel,** Inverse invariant theory and Steenrod operations, 2000

691 **Bruce Hughes and Stratos Prassidis,** Control and relaxation over the circle, 2000

690 **Robert Rumely, Chi Fong Lau, and Robert Varley,** Existence of the sectional capacity, 2000

689 **M. A. Dickmann and F. Miraglia,** Special groups: Boolean-theoretic methods in the theory of quadratic forms, 2000

688 **Piotr Hajłasz and Pekka Koskela,** Sobolev met Poincaré, 2000

687 **Guy David and Stephen Semmes,** Uniform rectifiability and quasiminimizing sets of arbitrary codimension, 2000

686 **L. Gaunce Lewis, Jr.,** Splitting theorems for certain equivariant spectra, 2000

685 **Jean-Luc Joly, Guy Metivier, and Jeffrey Rauch,** Caustics for dissipative semilinear oscillations, 2000

684 **Harvey I. Blau, Bangteng Xu, Z. Arad, E. Fisman, V. Miloslavsky, and M. Muzychuk,** Homogeneous integral table algebras of degree three: A trilogy, 2000

683 **Serge Bouc,** Non-additive exact functors and tensor induction for Mackey functors, 2000

682 **Martin Majewski,** ational homotopical models and uniqueness, 2000

681 **David P. Blecher, Paul S. Muhly, and Vern I. Paulsen,** Categories of operator modules (Morita equivalence and projective modules, 2000

680 **Joachim Zacharias,** Continuous tensor products and Arveson's spectral C^*-algebras, 2000

679 **Y. A. Abramovich and A. K. Kitover,** Inverses of disjointness preserving operators, 2000

678 **Wilhelm Stannat,** The theory of generalized Dirichlet forms and its applications in analysis and stochastics, 1999

(Continued in the back of this publication)

Quantum Linear Groups and Representations of $GL_n(\mathbb{F}_q)$

MEMOIRS
of the
American Mathematical Society

Number 706

Quantum Linear Groups and Representations of $GL_n(\mathbb{F}_q)$

Jonathan Brundan
Richard Dipper
Alexander Kleshchev

January 2001 • Volume 149 • Number 706 (first of 4 numbers) • ISSN 0065-9266

American Mathematical Society
Providence, Rhode Island

1991 *Mathematics Subject Classification.*
Primary 20C20, 20C33, 20G05, 17B37.

Library of Congress Cataloging-in-Publication Data

Brundan, Jonathan, 1970–
Quantum linear groups and representations of GLn(Fq) / Jonathan Brundan, Richard Dipper, Alexander Kleshchev.
p. cm. — (Memoirs of the American Mathematical Society, ISSN 0065-9266 ; no. 706)
"January 2001, Volume 149, Number 706 (first of 4 numbers)."
Includes bibliographical references.
ISBN 0-8218-2616-6 (alk. paper)
1. Linear algebraic groups. 2. Representations of groups. 3. Group schemes (Mathematics) I. Dipper, Richard. II. Kleshchëv, A. S. (Aleksandr Sergeevich) III. Title. IV. Series.
QA3 .A57 no. 706
[QA179]
510 s—dc21
[512′.2] 00-046918

Memoirs of the American Mathematical Society

This journal is devoted entirely to research in pure and applied mathematics.

Subscription information. The 2001 subscription begins with volume 149 and consists of six mailings, each containing one or more numbers. Subscription prices for 2001 are $494 list, $395 institutional member. A late charge of 10% of the subscription price will be imposed on orders received from nonmembers after January 1 of the subscription year. Subscribers outside the United States and India must pay a postage surcharge of $31; subscribers in India must pay a postage surcharge of $43. Expedited delivery to destinations in North America $35; elsewhere $130. Each number may be ordered separately; *please specify number* when ordering an individual number. For prices and titles of recently released numbers, see the New Publications sections of the *Notices of the American Mathematical Society.*

Back number information. For back issues see the *AMS Catalog of Publications.*

Subscriptions and orders should be addressed to the American Mathematical Society, P. O. Box 845904, Boston, MA 02284-5904. *All orders must be accompanied by payment.* Other correspondence should be addressed to Box 6248, Providence, RI 02940-6248.

Memoirs of the American Mathematical Society is published bimonthly (each volume consisting usually of more than one number) by the American Mathematical Society at 201 Charles Street, Providence, RI 02904-2294. Periodicals postage paid at Providence, RI. Postmaster: Send address changes to Memoirs, American Mathematical Society, P. O. Box 6248, Providence, RI 02940-6248.

This publication is indexed in *Science Citation Index®*, *SciSearch®*, *Research Alert®*, *CompuMath Citation Index®*, *Current Contents®/Physical, Chemical & Earth Sciences.*
Printed in the United States of America.

∞ The paper used in this book is acid-free and falls within the guidelines established to ensure permanence and durability.
Visit the AMS home page at URL: `http://www.ams.org/`

10 9 8 7 6 5 4 3 2 1 06 05 04 03 02 01

Contents

Abstract

We give a self-contained account of the results originating in the work of James and the second author in the 1980s relating the representation theory of $GL_n(\mathbb{F}_q)$ over fields of characteristic coprime to q to the representation theory of "quantum GL_n" at roots of unity.

The new treatment allows us to extend the theory in several directions. First, we prove a precise functorial connection between the operations of tensor product in quantum GL_n and Harish-Chandra induction in finite GL_n. This allows us to obtain a version of the recent Morita theorem of Cline, Parshall and Scott valid in addition for p-singular classes.

From that we obtain simplified treatments of various basic known facts, such as the computation of decomposition numbers and blocks of $GL_n(\mathbb{F}_q)$ from knowledge of the same for the quantum group, and the non-defining analogue of Steinberg's tensor product theorem. We also easily obtain a new double centralizer property between $GL_n(\mathbb{F}_q)$ and quantum GL_n, generalizing a result of Takeuchi.

Finally, we apply the theory to study the affine general linear group, following ideas of Zelevinsky in characteristic zero. We prove results that can be regarded as the modular analogues of Zelevinsky's and Thoma's branching rules. Using these, we obtain a new dimension formula for the irreducible cross-characteristic representations of $GL_n(\mathbb{F}_q)$, expressing their dimensions in terms of the characters of irreducible modules over the quantum group.

Key words and phrases: general linear groups, quantum linear groups, modular representation theory, quasihereditary algebras.

---

[1]Received by the editor May 3, 1999.

[2]1991 subject classification: 20C20, 20C33, 20G05, 17B37.

[3]Second author partially supported by the European TMR programme "Algebraic Lie Representations" (EC Network Contract No. ERB FMRX-CT97/0100).

[4]First and third authors partially supported by the NSF (grant nos DMS-9801442 and DMS-9600124).

Introduction

This article is a contribution to the study of the modular representation theory of the finite general linear group $GL_n(\mathbb{F}_q)$ over a field F of characteristic p coprime to q. We have attempted in the first place to give a self-contained account of the results originating in the work of James and the second author [D_1, D_2, DJ_2, DJ_3, J_1, J_2] in the 1980s relating representation theory of $GL_n(\mathbb{F}_q)$ to representation theory of "quantum GL_n" at roots of unity. Since that time, there have been a number of conceptual simplifications to the theory, e.g. in [CPS_3, D_3, D_4, DDu_1, DDu_2, HL_2, T], which we have incorporated in the present approach from the outset. We mention above all the centrally important Morita theorem of Cline, Parshall and Scott from [CPS_3, §9]. We will reprove this Morita theorem in the present article in a self-contained way, thus leading us to a new and independent approach to the results of [D_1, D_2, DJ_2, DJ_3, J_1, J_2] assumed in the Cline-Parshall-Scott argument. Along the way, we make many of these results more precise or more functorial, which is essential in order to prove the new results of the article described further below.

Our point of view has been to deduce as much as possible of the modular theory from standard, often purely character theoretic results in the characteristic zero theory of $GL_n(\mathbb{F}_q)$, combined with knowledge of the highest weight representation theory of quantum linear groups. For the former, we have adopted the point of view of the Deligne-Lusztig theory, as described for $GL_n(\mathbb{F}_q)$ by Fong and Srinivasan [FS], supplemented by various other basic results most of which can be found in Carter's book [C]; we also appeal to the result of [DDu_1, §5],[HL_2] showing that Harish-Chandra induction is independent of the choice of parabolic subgroup, and the basic result of block theory proved in [BM] (also originally proved in [FS] though we do not use the full block classfication of Fong and Srinivasan). For quantum linear groups, we have followed the treatment by Parshall and Wang [PW] wherever possible, as well as [Cl, JM, DDo, Do_7] for various additional results.

We now summarize the main steps in the development, so that we can describe the new results of the article in more detail. We restrict our attention in this introduction to the case of unipotent representations, though there is no such restriction in the main body of the article. So let F be an algebraically closed field of characteristic p coprime to q, and let $G_n = GL_n(\mathbb{F}_q)$. Let M denote the FG_n-module arising from the permutation representation of G_n on the cosets of a Borel subgroup. Following the idea of Cline, Parshall and Scott, we introduce the *cuspidal algebra*, which in the unipotent case is the quotient algebra $C_n := FG_n / \operatorname{ann}_{FG_n}(M)$. The cuspidal algebra C_n is actually a quotient algebra just of the "unipotent block" B_n of FG_n, by which we mean the union of blocks of the algebra FG_n corresponding to the irreducible constituents of the module M.

Fixing some arbitrary $m \geq n$, we let $S_{m,n}$ denote the q-Schur algebra. Following [Du], this can be viewed as the quotient algebra $S_{m,n} := U_m / \operatorname{ann}_{U_m}(V^{\otimes n})$ of Lusztig's divided power version from [L_2] of the quantized enveloping algebra $U_m := U_q(\mathfrak{gl}_m)$ (specialized over F at root of unity $v = q^{1/2}$), where $V^{\otimes n}$ is the n-fold tensor power of the natural m-dimensional U_m-module. (Actually, in the main body of the article, we work with the quantized coordinate ring rather than with U_m.) At the heart of the theory is an explicit Morita equivalence (see (3.5a)):

$$\beta_{m,n} : \operatorname{mod}(S_{m,n}) \to \operatorname{mod}(C_n).$$

The existence of this Morita equivalence was originally proved (in the unipotent case only) by Takeuchi [T], but we follow the quite different strategy of Cline, Parshall and Scott from [CPS_3, §9] for its construction. The idea is to exhibit an explicit projective generator for $\operatorname{mod}(C_n)$ with endomorphism algebra isomorphic to $S_{m,n}$. The projective generator used is roughly speaking the direct sum of the modules obtained by Harish-Chandra induction from all *Steinberg representations* of all standard Levi subgroups of G_n (the Steinberg representation does not in general remain irreducible on reduction modulo p so we refer the reader to §3.3 for the precise definition we use for this in characteristic p).

Write $L_m(\lambda)$ (resp. $\Delta_m(\lambda)$) for the irreducible (resp. standard or "Weyl") U_m-module of highest weight λ, where λ is a partition of height at most m. If λ is a partition of n, these modules factor through the quotient $S_{m,n}$, so we obtain the C_n-modules:

$$L(1, \lambda) := \beta_{m,n}(L_m(\lambda')),$$
$$\Delta(1, \lambda) := \beta_{m,n}(\Delta_m(\lambda')),$$

where λ' denotes the transpose partition. Since $\beta_{m,n}$ is a Morita equivalence, the modules $\{L(1, \lambda) \mid \lambda \vdash n\}$ give a complete set of non-isomorphic irreducible C_n-modules. It turns out moreover that their inflations to the unipotent block B_n give a complete set of non-isomorphic irreducible B_n-modules, while the *standard module* $\Delta(1, \lambda)$ is a reduction modulo p of an irreducible $\mathbb{C}G_n$-module affording the irreducible unipotent character χ_λ parametrized by the partition λ (cf. Theorem 4.1c, (4.4b)).

However, the unipotent block B_n has more irreducible characters in characteristic zero than just the characters $\{\chi_\lambda \mid \lambda \vdash n\}$, that is, one does not obtain complete information about the decomposition matrix of the unipotent block from the results described so far (only a square submatrix). To understand these additional irreducible characters in terms of the q-Schur algebra, our approach is to first prove an extension of the above Morita theorem to *arbitrary* elements $\sigma \in \bar{\mathbb{F}}_q^\times$ with p'-part equal to 1 (so the original unipotent case is then the special case that $\sigma = 1$).

The key new result in our proof of this extension relates the tensor product operator on U_m-modules to the Harish-Chandra induction operator $\diamond$ (more usually denoted by $\circ$) on GL-modules. To describe this, first recall that there is a natural notion of tensor product of two U_m-modules coming from the comultiplication of the Hopf algebra U_m. This means that for $n_1 + n_2 = n$, the tensor product $M_1 \otimes M_2$ of an S_{m,n_1}-module with an S_{m,n_2}-module can be viewed as an $S_{m,n}$-module in a natural way. In other words, there is a bifunctor

$$? \otimes ?' : \text{mod}(S_{m,n_1}) \times \text{mod}(S_{m,n_2}) \to \text{mod}(S_{m,n}).$$

Also Harish-Chandra induction gives us a bifunctor

$$? \diamond ?' : \text{mod}(FG_{n_1}) \times \text{mod}(FG_{n_2}) \to \text{mod}(FG_n).$$

We show in Theorem 4.2a that these two bifunctors correspond under the Morita equivalence, i.e. that the bifunctors $(\beta_{m,n_1}?) \diamond (\beta_{m,n_2}?')$ and $\beta_{m,n}(\,? \otimes ?'\,)$ from $\text{mod}(S_{m,n_1}) \times \text{mod}(S_{m,n_2})$ to $\text{mod}(FG_n)$ are isomorphic. In order to prove this, we first prove a q-analogue of the main result of [BK$_1$, §2] concerning polynomial induction in quantum linear groups, see §§1.4–1.5.

Using the p-singular generalization of the Morita theorem, one obtains additional standard FG_n-modules of the form $\Delta(\sigma, \lambda)$, for $\sigma \in \bar{\mathbb{F}}_q^\times$ of degree d over $\mathbb{F}_q$ with p'-part equal to 1, and $\lambda \vdash k$ where $n = kd$. If the image of q is a primitive ℓ-th root of unity in $F^\times$, there is an integer $r \geq 0$ so that $d = \ell p^r$. We show that the module $\Delta(\sigma, \lambda)$ can be realized alternatively under the original (unipotent) Morita equivalence as

$$\Delta(\sigma, \lambda) = \beta_{m,n}(\Delta_m(\lambda')^{[r]}),$$

where $M^{[r]}$ denotes the $S_{m,n}$-module obtained by taking the rth Frobenius twist of M (cf. Theorem 4.3d). Finally, to construct a general standard module of B_n, i.e. a module equal to the reduction modulo p of an *arbitrary* irreducible complex character of B_n, one needs to consider modules of the form

$$\Delta(\sigma_1, \lambda_1) \diamond \dots \diamond \Delta(\sigma_a, \lambda_a)$$

where the σ_i are non-conjugate elements of $\bar{\mathbb{F}}_q^\times$ with p'-part equal to 1 and the λ_i are partitions with $\sum_i |\lambda_i| \deg(\sigma_i) = n$ (cf. (4.4a)). In other words, the standard modules for the unipotent block B_n are lifts of C_n-modules which correspond under the Morita equivalence to Frobenius-twisted tensor products of standard modules for the quantum group of the form

$$\Delta_m(\lambda_1')^{[r_1]} \otimes \dots \otimes \Delta_m(\lambda_a')^{[r_a]}$$

(cf. Lemma 4.4c).

We then easily obtain alternative, functorial proofs of the main results of [FS, J$_2$, DJ$_3$], such as a description of the blocks of the algebra B_n (Theorem 4.4g) and an explicit formula for the decomposition numbers in terms of familiar (but unknown!) decomposition numbers and modular Littlewood-Richardson coefficients for the quantum algebra U_m (Theorem 4.4d). There is also a non-defining characteristic analogue of Steinberg's tensor product theorem (Theorem 4.3e) for irreducible modules obtained originally by Du and the second author [DDu$_2$], which now follows immediately from the tensor product theorem for U_m and the other results described so far.

We mentioned earlier the work of Takeuchi. From our point of view, Takeuchi's double centralizer property from [T] follows as an elementary consequence of a more general double centralizer property, which we regard as the non-defining characteristic analogue of Donkin-Howe duality [Do$_5$]. This has a particularly simple formulation in the unipotent case. Let T be the permutation representation of FG_n on the set of all m-step flags

$$0 = f_0 \subseteq f_1 \subseteq \cdots \subseteq f_m = W_n,$$

where $W_n = \mathbb{F}_q^n$ is the natural representation of G_n. It turns out that the action of FG_n on T factors through the quotient C_n of FG_n, so that T is a C_n-module in a natural way, and moreover, the endomorphism algebra $\mathrm{End}_{C_n}(T)$ is isomorphic to $S_{m,n}$ (see Theorem 3.4a). This was originally proved in [DJ$_3$, Theorem 2.24] (see also the construction in [BLM]). In Theorem 4.5e, we prove the following double centralizer property:

$$\mathrm{End}_{S_{m,n}}(T) = C_n.$$

This result is very natural from the point of view of quasi-hereditary algebras; since C_n is Morita equivalent to $S_{m,n}$, and the latter is a quasi-hereditary algebra according to [PW], C_n is itself quasi-hereditary. The permutation representation T is in fact a full tilting module for the quasi-hereditary algebra C_n in the sense of Ringel [R], so that $S_{m,n}$ is a Ringel dual of C_n (cf. Theorem 4.5d).

The final results of the article are concerned with the affine general linear group $H_n := AGL_n(\mathbb{F}_q)$, that is, the semi-direct product G_nW_n of the elementary Abelian group W_n by G_n. Our results here were motivated by the ideas of Zelevinsky [Z, §13] in characteristic 0. Corresponding to the natural embedding $U_m \hookrightarrow U_{m+1}$ there is a Levi subalgebra of the q-Schur algebra $S_{m+1,n}$ which we denote by

$$S_{m,\leq n} \cong \bigoplus_{j=0}^{n} S_{m,j}.$$

We define an affine analogue D_n of the cuspidal algebra C_n, namely, the algebra $FH_n/\operatorname{ann}_{FH_n}(N)$ where N is the permutation representation of FH_n on the cosets of a Borel subgroup of $G_n \subset H_n$ ("affine flags"). There is an explicit Morita equivalence (cf. (5.3d)):

$$\beta_{m,\leq n} : \mathrm{mod}(S_{m,\leq n}) \to \mathrm{mod}(D_n).$$

Regarding $\beta_{m+1,n}$ (resp. $\beta_{m,\leq n}$) now as a functor from $\mathrm{mod}(S_{m+1,n})$ to FG_n (resp. from $\mathrm{mod}(S_{m,\leq n})$ to FH_n), our main result on the affine linear group (Theorem 5.4c) shows that there is an isomorphism of functors:

$$\operatorname{ind}_{G_n}^{H_n} \circ \beta_{m+1,n} \cong \beta_{m,\le n} \circ \operatorname{res}^{S_{m+1,n}}_{S_{m,\le n}}.$$

This allows us to relate the problem of decomposing an induced module of the form $\operatorname{ind}_{G_n}^{H_n} L$ for an irreducible FG_n-module L to the problem of decomposing the restriction of an irreducible U_{m+1}-module to U_m. From this, we also explain how to calculate the composition multiplicities of the restriction $\operatorname{res}^{G_n}_{H_{n-1}} L$ from knowledge of the composition multiplicities of restrictions of irreducibles from U_m to U_{m-1}, a result which we regard as a modular analogue of Zelevinsky's branching rule from [Z, Theorem 13.5] (in turn, this is really an extension of Thoma's branching rule [Th]).

Finally, there is a well-known hook formula (2.3.3) for the degrees of the irreducible complex characters of G_n, but in positive characteristic remarkably little is known about the dimensions of the irreducible FG_n-modules (see e.g. [GT]). As an application of our branching rule for the affine linear group, we obtain a dimension formula (Theorem 5.5d) for the irreducible FG_n-modules in terms of (unknown!) characters of irreducible U_m-modules. To state this for unipotent representations, define the polynomial $S_\lambda(t) \in \mathbb{Z}[t]$ for a partition $\lambda = (l_1 \geq l_2 \geq \dots \geq l_h > 0)$ of height h by

$$S_\lambda(t) := \sum_{(m_1,\dots,m_h)} \left[\prod_{i=1}^{n} (t^i - 1) \Big/ \prod_{i=1}^{h} (t^{m_1+\dots+m_i} - 1) \right],$$

where the sum is over all h-tuples $(m_1, \dots, m_h)$ that can be obtained by reordering the non-zero parts $l_1, \dots, l_h$ of λ. Then, we show that

$$\dim L(1, \lambda') = \sum_{\mu \vdash n} m_{\lambda,\mu} S_\mu(q)$$

where $m_{\lambda,\mu}$ is the weight multiplicity of the weight μ in the irreducible U_m-module $L_m(\lambda)$ of highest weight λ. Even though the latter weight multiplicities $m_{\lambda,\mu}$ are unknown in general, we expect that this result can be used to improve the known bounds for the low dimensional representations of $GL_n(\mathbb{F}_q)$ in cross characteristics of [GT].

We now give a brief description of the layout of the article. First, we mention that there is a very brief guide to some of our non-standard notation for the expert reader, immediately following this introduction. Then, there are five further chapters. The first gives a rapid review of the basic results concerning quantum linear groups and the q-Schur algebra that we need, as well as proving the q-analogue of the results on polynomial induction from [BK$_1$]. Chapter 2 gives a similar review of the basic results from the characteristic zero theory of $GL_n(\mathbb{F}_q)$.

The core of the theory is explained in chapter 3, where in particular we prove the most important special case of the Morita theorem and use it to introduce the various basic modules. Note that throughout chapter 3 (and the early part of chapter 4), we make certain standing assumptions (A1) and (A2) on the element $\sigma \in \bar{\mathbb{F}}_q^\times$, which we only know *a priori* are satisfied if σ is p-regular. Only in chapter 4 are we able to show that these assumptions are satisfied in general, so that the earlier results are true without restriction, when we prove the extension of the Morita theorem to p-singular classes. Also in chapter 4 we discuss an integral version of the Morita theorem allowing us to understand base change and prove the result relating tensor products to Harish-Chandra induction mentioned earlier. We then give the applications to reprove the results on decomposition numbers of [DJ3] and the non-defining tensor product theorem of [DDu2], and end by proving the double centralizer property. Chapter 5 contains the results on the affine linear group.

Acknowledgements. We would like to thank the organizers of the "Algebraic Representation Theory" conference in Aarhus in 1998, where part of the research for this article was carried out. We also thank S. Donkin for communicating the proof of Lemma 4.5a to us, and S. König and R. Rouquier for help with the literature in various places.

Notation

Conventions:

Unless we specify otherwise, F is a commutative, unital ring (from chapter 2 onwards F will always be an algebraically closed field of characteristic p) and $\otimes$ denotes tensor product over F.

Given an algebra C over F, a C-module always means a *left* C-module that is *finitely generated* over F unless we explicitly say otherwise. We will write $\mathrm{mod}(C)$ for the category of all such finitely generated left C-modules. Given a C-module M, the endomorphism algebra $\mathrm{End}_C(M)$ is always assumed to act on M on the right.

For an F-coalgebra A, an A-comodule will always mean a *right* A-comodule, not necessarily finitely generated. So an A-comodule M is an F-module together with a structure map $\tau : M \to M \otimes A$ satisfying the usual axioms, see Sweedler [Sw]. We write $\mathrm{comod}(A)$ for the category of all such right A-comodules.

Notation overview:

G_n The general linear group $GL_n(\mathbb{F}_q)$ over the finite field $\mathbb{F}_q$ of q elements, where q is a prime power not divisible by p.

h, k, d Positive integers such that $h \geq k$ and $n = kd$.

H_k The Hecke algebra $H_{F,q^d}(\Sigma_k)$ associated to the symmetric group Σ_k, with standard basis $\{T_w \mid w \in \Sigma_k\}$ (p.10).

$S_{h,k}$ The q^d-Schur algebra $S_{F,q^d}(h,k)$, which is the quotient of Lusztig's divided power version of the quantum algebra $U_{F,q^{d/2}}(\mathfrak{gl}_h)$ (over F at root of unity $q^{d/2}$) under its representation on the kth tensor power of its h-dimensional natural module (p.12).

σ An element of $\bar{\mathbb{F}}_q^\times$ of degree d over $\mathbb{F}_q$, with associated companion matrix $(\sigma) \in G_d$ (p.29).

$M(\sigma)$ The irreducible, cuspidal FG_d-module associated to σ (p.37).

$M^k(\sigma)$ The left FG_n-module obtained by Harish-Chandra induction from the outer tensor product $M(\sigma) \boxtimes \cdots \boxtimes M(\sigma)$ (k times); e.g. in the unipotent case, $M^n(1)$ is the permutation representation of FG_n on cosets of a Borel subgroup. There is an explicit, fixed isomorphism $\mathrm{End}_{FG_n}(M^k(\sigma)) \cong H_k$ (p.38).

$C_{\sigma,k}$ The *cuspidal algebra*, namely, the image of FG_n under its representation on $M^k(\sigma)$ (p.51).

$Z^k(\sigma)$ The largest submodule of $M^k(\sigma)$ on which each $T_w \in H_k$ acts as $(-1)^{\ell(w)}$; e.g. in the unipotent case, $Z^n(1)$ is a modular reduction of the Steinberg module (p.54).

$\Lambda^k(\sigma)$ The irreducible FG_n-module $M^k(\sigma)(\sum_{w\in\Sigma_k} T_w)$; e.g. in the unipotent case, $\Lambda^n(1)$ is the trivial representation of FG_n (p.55).

$\dot{Z}^\nu(\sigma)$ For $\nu = (k_1, \ldots, k_a) \vDash k$, $\dot{Z}^\nu(\sigma)$ is the FG_n-module obtained by Harish-Chandra induction from $Z^\nu(\sigma) = Z^{k_1}(\sigma) \boxtimes \cdots \boxtimes Z^{k_a}(\sigma)$. Regarded as a $C_{\sigma,k}$-module, $\dot{Z}^\nu(\sigma)$ is *projective* and $\bigoplus_\nu \dot{Z}^\nu(\sigma)$, summing over all compositions ν of k with at most h rows, is a projective generator for $C_{\sigma,k}$. The endomorphism algebra $\mathrm{End}_{C_{\sigma,k}}(\bigoplus_\nu \dot{Z}^\nu(\sigma))$ is isomorphic to $S_{h,k}$, so $S_{h,k}$ and $C_{\sigma,k}$ are Morita equivalent (p.57).

$\dot{\Lambda}^\nu(\sigma)$ For $\nu = (k_1, \ldots, k_a) \vDash k$, $\dot{\Lambda}^\nu(\sigma)$ is the FG_n-module Harish-Chandra induced from $\Lambda^\nu(\sigma) = \Lambda^{k_1}(\sigma) \boxtimes \cdots \boxtimes \Lambda^{k_a}(\sigma)$. The endomorphism algebra $\mathrm{End}_{C_{\sigma,k}}(\bigoplus_\nu \dot{\Lambda}^\nu(\sigma))$, summing over all compositions ν of k with at most h rows, is again isomorphic to $S_{h,k}$. In addition, the double centralizer property holds, that is, $\mathrm{End}_{S_{h,k}}(\bigoplus_\nu \dot{\Lambda}^\nu(\sigma))$ is isomorphic to $C_{\sigma,k}$. From this point of view, $\bigoplus_\nu \dot{\Lambda}^\nu(\sigma)$ is a full tilting module for the quasi-hereditary algebra $C_{\sigma,k}$, and $S_{h,k}$ and $C_{\sigma,k}$ are *Ringel duals* (p.57).

$\alpha_{\sigma,h,k}$ The functor $\mathrm{Hom}_{C_{\sigma,k}}(\bigoplus_\nu \dot{Z}^\nu(\sigma), ?)$ which yields the equivalence of categories between $\mathrm{mod}(C_{\sigma,k})$ and $\mathrm{mod}(S_{h,k})$ (p.63).

$\beta_{\sigma,h,k}$ The functor $(\bigoplus_\nu \dot{Z}^\nu(\sigma)) \otimes_{S_{h,k}} ?$ which is inverse to the Morita equivalence $\alpha_{\sigma,h,k}$ (p.63).

$\chi_{\sigma,\lambda}$ For $F = \mathbb{C}$, the ordinary irreducible characters arising as constituents of $M^k(\sigma)$ are the characters $\{\chi_{\sigma,\lambda} \mid \lambda \vdash k\}$ (p.36).

$\Delta(\sigma,\lambda)$ The modules $\{\Delta(\sigma,\lambda) \mid \lambda \vdash k\}$ are the standard $C_{\sigma,k}$-modules (in the sense of quasi-hereditary algebras). The standard module $\Delta(\sigma,\lambda)$ is a modular reduction of a $\mathbb{C}G_n$-module affording the character $\chi_{\sigma,\lambda}$, and corresponds under the Morita equivalence to the standard $S_{h,k}$-module of highest weight λ' (transpose partition). Explicitly, $\Delta(\sigma,\lambda)$ can be defined as the image of any non-zero element of the one dimensional space $\mathrm{Hom}_{FG_n}(\dot{Z}^{\lambda'}(\sigma), \dot{\Lambda}^\lambda(\sigma))$ (p.63).

$L(\sigma,\lambda)$ The simple head of $\Delta(\sigma,\lambda)$. The $\{L(\sigma,\lambda) | \lambda \vdash k\}$ give a complete set of non-isomorphic irreducible $C_{\sigma,k}$-modules (p.63).

$(s, \underline{\lambda})$ An arbitrary semisimple element $s \in G_n$ can be written up to conjugacy in block-diagonal form as $s = \mathrm{diag}((\sigma_1)^{k_1} \ldots (\sigma_a)^{k_a})$ for non-conjugate elements $\sigma_1, \ldots, \sigma_a$ of $\bar{\mathbb{F}}_q^\times$ and $k_1, \ldots, k_a \geq 1$. For such an s, fix also a multi-partition $\underline{\lambda} = (\lambda_1, \ldots, \lambda_a)$ with each $\lambda_i \vdash k_i$ (p.30).

$\chi_{s,\underline{\lambda}}$ The ordinary irreducible character obtained from $\chi_{\sigma_1,\lambda_1} \cdots \chi_{\sigma_a,\lambda_a}$ by Harish-Chandra induction (p.35).

$\Delta(s,\underline{\lambda})$ The module obtained by Harish-Chandra induction from $\Delta(\sigma_1,\lambda_1) \boxtimes \cdots \boxtimes \Delta(\sigma_a,\lambda_a)$, and a modular reduction of a $\mathbb{C}G_n$-module affording the character $\chi_{s,\underline{\lambda}}$ (p.79).

$L(s,\underline{\lambda})$ Suppose in addition that s is p-regular. Then, $L(s,\underline{\lambda})$ is the module obtained by Harish-Chandra induction from $L(\sigma_1,\lambda_1) \boxtimes \cdots \boxtimes L(\sigma_a,\lambda_a)$, and is isomorphic to the simple head of $\Delta(s,\underline{\lambda})$. All irreducible FG_n-modules have this form (p.79).

Chapter 1

Quantum linear groups and polynomial induction

In this first chapter, we collect together all the known results about quantum linear groups and the q-Schur algebra that we will need later. Then we prove some new results about polynomial induction in quantum linear groups, generalizing results of [BK$_1$, §2] in the classical setting.

1.1. Symmetric groups and Hecke algebras

Fix an integer $k \geq 1$. We write $\nu \vDash k$ if ν is a *composition of* k, so that $\nu = (k_1, k_2, \dots)$ for non-negative integers $k_1, k_2, \dots$ such that $k_1 + k_2 + \cdots = k$. If in addition $k_1 \geq k_2 \geq \dots$, then ν is a *partition of* k, written $\nu \vdash k$. The *height* $h(\nu)$ of a composition $\nu = (k_1, k_2, \dots) \vDash k$ is the smallest integer $a \geq 1$ such that $k_{a+1} = k_{a+2} = \cdots = 0$. For $\lambda \vdash k$, write λ' for the *transpose partition*, i.e. the partition whose Young diagram is obtained by reflecting the Young diagram of λ in the main diagonal. Let $\leq$ denote the usual dominance order on the set of all compositions of k, namely, $\mu = (m_1, m_2, \dots) \leq \lambda = (l_1, l_2, \dots)$ if and only if $\sum_{i=1}^{j} m_i \leq \sum_{i=1}^{j} l_i$ for all $j \geq 1$. We also write $\mu \geq \lambda$ if $\lambda \leq \mu$, and $\mu < \lambda$ if $\mu \leq \lambda$ but $\mu \neq \lambda$.

We write Σ_k for the symmetric group on k letters. For $w \in \Sigma_k$, $\ell(w)$ is the *length* of w, that is, the minimal number ℓ of basic transpositions $s_1, \dots, s_\ell$ such that $w = s_1 s_2 \dots s_\ell$. For $\nu = (k_1, \dots, k_a) \vDash k$, Σ_ν denotes the *Young subgroup* of Σ_k isomorphic to $\Sigma_{k_1} \times \cdots \times \Sigma_{k_a}$.

For $\lambda, \mu \vDash k$, the set D_λ (resp. D_μ^{-1}) of elements of Σ_k which are of minimal length in their Σ_k/Σ_λ-coset (resp. their $\Sigma_\mu\backslash\Sigma_k$-coset) gives a set of distinguished Σ_k/Σ_λ-coset representatives (resp. $\Sigma_\mu\backslash\Sigma_k$-coset representatives). We set $D_{\mu,\lambda} = D_\mu^{-1} \cap D_\lambda$, to obtain a set of distinguished $\Sigma_\mu\backslash\Sigma_k/\Sigma_\lambda$-coset representatives. Moreover, if both Σ_μ and Σ_λ are subgroups of Σ_ν for some $\nu \vDash k$, the set $D_{\mu,\lambda}^\nu = D_{\mu,\lambda} \cap \Sigma_\nu$ is a set of $\Sigma_\mu\backslash\Sigma_\nu/\Sigma_\lambda$-coset representatives.

We will freely use well-known properties of these distinguished double coset representatives, all of which can be found in [C, §2.7] or [DJ$_1$, §1]. We note in particular the following:

(1.1a) *Given* $\lambda, \mu \models k$ *and any* $w \in D_{\mu,\lambda}$, $\Sigma_\mu \cap {}^w\Sigma_\lambda$ *is also a Young subgroup of* Σ_k.

Now suppose that F is an arbitrary commutative ring and $q \in F$ is arbitrary. The *Iwahori-Hecke algebra* $H_{F,q}(\Sigma_k)$ associated to the symmetric group Σ_k over F at parameter q is a certain F-free F-algebra with basis $\{T_w \mid w \in \Sigma_k\}$ and satisfying the relations

$$T_wT_s = \begin{cases} T_{ws} & \text{if } \ell(ws) = \ell(w)+1, \\ qT_{ws} + (q-1)T_w & \text{if } \ell(ws) = \ell(w) - 1 \end{cases}$$

for all $w \in \Sigma_k$ and all basic transpositions $s \in \Sigma_k$. For an indeterminate t, regard the ring F as a $\mathbb{Z}[t]$-module by letting t act on F by multiplication by q. We have that:

$$H_{F,q}(\Sigma_k) \cong F \otimes_{\mathbb{Z}[t]} H_{\mathbb{Z}[t],t}(\Sigma_k), \tag{1.1.1}$$

the isomorphism being the obvious one sending the basis element $1 \otimes T_w$ of $F \otimes_{\mathbb{Z}[t]} H_{\mathbb{Z}[t],t}(\Sigma_k)$ to the corresponding basis element T_w of $H_{F,q}(\Sigma_k)$.

Write simply H_k for $H_{F,q}(\Sigma_k)$. There are two F-free H_k-modules of rank one, and we next recall their definitions. We let

$$x_k = \sum_{w \in \Sigma_k} T_w \qquad \text{and} \qquad y_k = (-1)^{\ell(w_0)} \sum_{w \in \Sigma_k} (-q)^{\ell(w_0)-\ell(w)} T_w$$

where $\ell(w_0) = \frac{1}{2}k(k-1)$. Note that our definition of y_k is different from the original definition in [DJ$_1$], but only up to a scalar; the present definition allows the basic results to be stated also if q is not a unit, and is more convenient in view of (1.1c). According to [DJ$_1$, pp.28–29]:

(1.1b) *For* $w \in \Sigma_k$, $T_wx_k = x_kT_w = q^{\ell(w)}x_k$ *and* $T_wy_k = y_kT_w = (-1)^{\ell(w)}y_k$.

The left H_k-modules $\mathcal{I}_{H_k}$ and $\mathcal{E}_{H_k}$, called the *trivial module* and the *sign module* respectively, can now be defined as the left ideals $\mathcal{I}_{H_k} = H_kx_k$ and $\mathcal{E}_{H_k} = H_ky_k$. We will also write $\mathcal{I}_{H_k}$ and $\mathcal{E}_{H_k}$ for the functions from H_k to F that arise from the action of H_k on these rank one modules. It will be important to know that if F is a field and $q \neq 0$, then the modules $\mathcal{I}_{H_k}$ and $\mathcal{E}_{H_k}$ are the *only* one dimensional H_k-modules (cf. [DJ$_1$, Lemma 3.1]).

The algebra H_k possesses an involutive automorphism $\#$ defined on generators by $T_s^\# = -T_s + q - 1$ for a basic transposition s. For any left H_k-module M, we let $M^\#$ denote the module which is equal to M as an F-module, but with new action defined by $h \cdot m = h^\# m$ for $m \in M, h \in H_k$. This is the analogue for the Hecke algebra of the module M tensored with the sign representation in the symmetric group setting. We record the elementary calculation, showing that $\#$ swaps the trivial and sign representations (e.g. see [T, Proposition 2.2]):

(1.1c) $x_k^{\#} = y_k$ *and* $y_k^{\#} = x_k$.

Let $\nu = (k_1, \ldots, k_a) \models k$. The above definitions generalize easily to the *Young subalgebra* $H_\nu = H_{F,q}(\Sigma_\nu)$ of H_k, which is the subalgebra spanned by $\{T_w \,|\, w \in \Sigma_\nu\}$. Identifying H_ν and $H_{k_1} \otimes \cdots \otimes H_{k_a}$ in the natural way, we let x_ν (resp. y_ν) denote the element $x_{k_1} \otimes \cdots \otimes x_{k_a}$ (resp. $y_{k_1} \otimes \cdots \otimes y_{k_a}$) of H_ν. The H_ν-modules $\mathcal{I}_{H_\nu} = H_\nu x_\nu$ and $\mathcal{E}_{H_\nu} = H_\nu y_\nu$ are then the trivial and sign representations of H_ν.

As in [DJ$_1$, §3], we define the *permutation module* M^ν and *signed permutation module* N^ν of H_k to be left ideals $H_k x_\nu$ and $H_k y_\nu$ respectively. By (1.1c), we have that $(M^\nu)^{\#} \cong N^\nu$. Moreover, H_k is a free right H_ν-module with basis $\{T_w | w \in D_\nu\}$. Therefore, M^ν can also be defined as the *induced module* $\mathrm{ind}_{H_\nu}^{H_k} \mathcal{I}_{H_\nu} = H_k \otimes_{H_\nu} \mathcal{I}_{H_\nu}$. Similarly, $N^\nu \cong \mathrm{ind}_{H_\nu}^{H_k} \mathcal{E}_{H_\nu} = H_k \otimes_{H_\nu} \mathcal{E}_{H_\nu}$.

Define u_λ to be the unique element of $D_{\lambda',\lambda}$ such that $\Sigma_{\lambda'} \cap {}^{u_\lambda}\Sigma_\lambda = \{1\}$. Note that u_λ is precisely the element denoted $w_{\lambda'}$ in [DJ$_4$, p.258]. We need the following fact proved in [DJ$_1$, Lemma 4.1]:

(1.1d) *For* $\lambda \vdash k$, *the space* $y_{\lambda'} H_k x_\lambda$ *is an* F*-free* F*-module of rank one, generated by the element* $y_{\lambda'} T_{u_\lambda} x_\lambda$.

The *Specht module* Sp^λ is the left ideal $Sp^\lambda = H_k y_{\lambda'} T_{u_\lambda} x_\lambda$. Observe that Sp^λ is both a submodule of M^λ and a quotient of $N^{\lambda'}$. As is well-known (see e.g. [DJ$_1$, Theorem 4.15]) if F is a field of characteristic 0 and q is a positive integer, H_k is a semisimple algebra and the Specht modules $\{Sp^\lambda \mid \lambda \vdash k\}$ give a complete set of non-isomorphic irreducible H_k-modules. In this case, we also have the following well-known characterization of Specht modules:

(1.1e) *For* F *a field of characteristic* 0 *and* q *a positive integer,* Sp^λ *is the unique irreducible* H_k*-module that is a constituent of both* $N^{\lambda'}$ *and* M^λ *(having multiplicity one in each).*

Of course, we can take the very special case with $F = \mathbb{Q}$ and $q = 1$. Then, $H_{\mathbb{Q},1}(\Sigma_k)$ is just the group algebra $\mathbb{Q}\Sigma_k$ of the symmetric group over $\mathbb{Q}$ and we see that the Specht modules $\{Sp^\lambda \mid \lambda \vdash k\}$ give a complete set of non-isomorphic irreducible $\mathbb{Q}\Sigma_k$-modules. We will write $X(\Sigma_k)$ for the character ring of Σ_k over $\mathbb{Q}$. For $\lambda \vdash k$, let

$$\phi_\lambda \in X(\Sigma_k) \tag{1.1.2}$$

denote the irreducible character of the symmetric group corresponding to the Specht module Sp^λ over $\mathbb{Q}$. Given in addition $\mu \vdash k$, we write $\phi_\lambda(\mu)$ for the value of the character ϕ_λ on any element of Σ_k of cycle-type μ.

1.2. The q-Schur algebra

Continue initially with F denoting an arbitrary ring and $q \in F$ being arbitrary. Fix $h \geq 1$. We write $\Lambda(h,k)$ for the set of all compositions $\nu = (k_1, \ldots, k_h) \models k$ of height at most h and $\Lambda^+(h,k)$ for the set of all *partitions* $\nu \in \Lambda(h,k)$. Let $\Lambda(h)$ be the set of all h-tuples of non-negative integers. Given $\lambda, \mu \in \Lambda(h)$, we write $\lambda + \mu$ for their coordinate-wise sum, and $c\lambda$ denotes $\lambda + \cdots + \lambda$ (c times). Identifying $\Lambda(h)$ with the union $\bigcup_{k \geq 0} \Lambda(h,k)$, let $\Lambda^+(h) = \bigcup_{k \geq 0} \Lambda^+(h,k) \subseteq \Lambda(h)$. We refer to the elements of $\Lambda^+(h)$ as *dominant weights.*

Following [DJ3, DJ4], we define the *q-Schur algebra* $S_{h,k} = S_{F,q}(h,k)$ to be the endomorphism algebra

$$\mathrm{End}_{H_k}\left(\bigoplus_{\nu \in \Lambda(h,k)} M^\nu \right),$$

writing endomorphisms commuting with the left action of H_k on the right. By convention, the algebra $S_{0,k} = S_{F,q}(0,k)$ is the trivial algebra F.

The q-Schur algebra is F-free and has a natural basis corresponding to certain double coset sums in H_k, which we now describe. Fix initially $\lambda, \mu \in \Lambda(h,k)$. For $u \in \Sigma_k$, we note that

$$\sum_{w \in \Sigma_\mu u \Sigma_\lambda} T_w = \sum_{w \in \Sigma_\mu u \Sigma_\lambda \cap D_\mu^{-1}} x_\mu T_w = \sum_{w \in \Sigma_\mu u \Sigma_\lambda \cap D_\lambda} T_w x_\lambda. \tag{1.2.1}$$

So, right multiplication in H_k by the element $\sum_{w \in \Sigma_\mu u \Sigma_\lambda \cap D_\mu^{-1}} T_w$ induces a well-defined homomorphism of left H_k-modules

$$\phi^u_{\mu,\lambda} : H_k x_\mu \to H_k x_\lambda.$$

Extending $\phi^u_{\mu,\lambda}$ to all of $\bigoplus_{\nu \in \Lambda(h,k)} M^\nu$ by letting it act as zero on M^ν for $\nu \neq \mu$, we obtain a well-defined element $\phi^u_{\mu,\lambda}$ of $S_{h,k}$. Now we can state the well-known result, proved originally in [DJ4, Theorem 1.4], under slightly more restrictive assumptions than here; see also [Ma, Theorem 4.8] for an argument valid in general.

(1.2a) *$S_{h,k}$ is F-free with basis $\{\phi^u_{\mu,\lambda} \mid \mu, \lambda \in \Lambda(h,k), u \in D_{\mu,\lambda}\}$.*

We refer to the basis for $S_{h,k}$ of (1.2a) as the *natural basis.* One shows easily using (1.2a) that $S_{h,k}$ behaves well under base change. To be precise, one has the analogue of (1.1.1):

$$S_{F,q}(h,k) \cong F \otimes_{\mathbb{Z}[t]} S_{\mathbb{Z}[t],t}(h,k)$$

where we are regarding F as a $\mathbb{Z}[t]$-algebra by letting t act on F as multiplication by q. We also note at this point the following well-known property, which is an immediate consequence of the definition of the natural basis of $S_{h,k}$ given above:

(1.2b) *For $h \geq k$, the F-linear map $\kappa : H_k \to S_{h,k}$, defined on a basis element T_w for $w \in \Sigma_k$ by $\kappa(T_w) = \phi^w_{(1^k),(1^k)}$, is a ring embedding.*

We could have chosen to define the q-Schur algebra $S_{h,k}$ equally well using the signed permutation module N^ν instead of M^ν, and we will later need this alternative point of view. So consider instead the algebra $\mathrm{End}_{H_k}\left(\bigoplus_{\lambda\in\Lambda(h,k)} N^\lambda\right)$, writing endomorphisms on the right again. Applying # to (1.2.1), we note that

$$\sum_{w\in\Sigma_\mu u\Sigma_\lambda} T_w^{\#} = \sum_{w\in\Sigma_\mu u\Sigma_\lambda\cap D_\mu^{-1}} y_\mu T_w^{\#} = \sum_{w\in\Sigma_\mu u\Sigma_\lambda\cap D_\lambda} T_w^{\#} y_\lambda. \tag{1.2.2}$$

Now, as in [DJ$_3$, p.26], the following fact follows easily:

(1.2c) *The algebras* $S_{h,k}$ *and* $\mathrm{End}_{H_k}(\bigoplus_{\lambda\in\Lambda(h,k)} N^\lambda)$ *are isomorphic, the natural basis element* $\phi^u_{\mu,\lambda}$ *of* $S_{h,k}$ *corresponding under the isomorphism to the endomorphism which is zero on* N^ν *for* $\nu\neq\mu$ *and sends* N^μ *into* N^λ *via the homomorphism induced by right multiplication in* H_k *by* $\sum_{w\in\Sigma_\lambda u\Sigma_\mu\cap D_\mu^{-1}} T_w^{\#}\in H_k$.

Henceforth, we assume that F is a field and that $q\in F$ is non-zero. Now we briefly recall some basic facts about the representation theory of the finite dimensional F-algebra $S_{h,k}$. The irreducible $S_{h,k}$-modules are parametrized by the set $\Lambda^+(h,k)$ of all partitions of k of height at most h. We will write $L_h(\lambda)$ for the irreducible $S_{h,k}$-module corresponding to $\lambda\in\Lambda^+(h,k)$ in the standard way. So $L_h(\lambda)$ is the unique irreducible $S_{h,k}$-module L with highest weight λ, that is, $\phi^1_{\lambda,\lambda}L\neq 0$ and $\phi^1_{\mu,\mu}L=0$ for all $\mu\in\Lambda(h,k)$ with $\mu\not\leq\lambda$.

Then, $S_{h,k}$ is a *quasi-hereditary algebra* with weight poset $(\Lambda^+(h,k),\leq)$, in the sense of Cline, Parshall and Scott [CPS$_2$]. This was first proved by Parshall and Wang [PW]; for other proofs see [Do$_6$, §4] (which follows the original homological proof from [Do$_3$, Do$_4$] of the classical analogue), or [Gr] or [Ma] (which are more combinatorial in nature). For recent accounts of the theory of quasi-hereditary algebras, see [Do$_7$, Appendix] or [KK].

In particular, we have associated to $\lambda\in\Lambda^+(h,k)$ (in a canonical way) the modules $\Delta_h(\lambda)$ and $\nabla_h(\lambda)$, which are the *standard* and *costandard* modules corresponding to $L_h(\lambda)$. We record the well-known properties:

(1.2d) $\Delta_h(\lambda)$ *(resp.* $\nabla_h(\lambda)$*) has simple head (resp. socle) isomorphic to* $L_h(\lambda)$*, and all other composition factors are of the form* $L_h(\mu)$ *with* $\mu<\lambda$.

In addition to being a quasi-hereditary algebra, the algebra $S_{h,k}$ possesses an anti-automorphism τ defined on the standard basis element $\phi^u_{\mu,\lambda}$ by $\tau(\phi^u_{\mu,\lambda})=\phi^{u^{-1}}_{\lambda,\mu}$ (see [DJ$_4$, Theorem 1.11]). Using this, we define the *contravariant dual* M^τ of an $S_{h,k}$-module M to be the dual vector space M^* with action defined by $(s.f)(m)=f(\tau(s)m)$ for all $s\in S_{h,k}, m\in M, f\in M^*$. This duality fixes the simple modules, that is, $L_h(\lambda)\cong L_h(\lambda)^\tau$ for all $\lambda\in\Lambda^+(h,k)$. We also note that $\Delta_h(\lambda)\cong\nabla_h(\lambda)^\tau$.

Given a left $S_{h,k}$-module M, we will write $\tilde{M}$ for the right $S_{h,k}$-module equal to M as a vector space with right action defined by $ms=\tau(s)m$ for $m\in M, s\in S_{h,k}$. In particular, this gives us modules $\tilde{L}_h(\lambda)$, $\tilde{\Delta}_h(\lambda)$ and $\tilde{\nabla}_h(\lambda)$ for each $\lambda\in\Lambda^+(h,k)$. We will use the following well-known result:

(1.2e) *$S_{h,k}$ has a filtration as an $(S_{h,k}, S_{h,k})$-bimodule with factors isomorphic to $\Delta_h(\lambda) \otimes \tilde{\Delta}_h(\lambda)$, each appearing precisely once for each $\lambda \in \Lambda^+(h,k)$ and ordered in any way refining the dominance order on partitions so that factors corresponding to most dominant λ appear at the bottom of the filtration.*

Actually, (1.2e) is a special case of a general property of quasi-hereditary algebras with an anti-automorphism τ fixing the simple modules as in the previous paragraph. It follows directly from the definition of quasi-hereditary algebra in terms of heredity ideals (see e.g. [CPS$_2$, p.92] or [KK, §1]) together with [DR, Statement 7] (one needs to observe that the module Ae of *loc. cit.* is a standard module and, by our assumption that τ fixes the simple modules, that $eA \cong \widetilde{Ae}$). For the filtration of (1.2e) in the classical case, see [Do$_3$, (3.2c)] and [Do$_4$, (1.5)].

1.3. Tensor products and Levi subalgebras

To describe further results about the q-Schur algebra, we need to relate it to the quantum linear group. Actually, we only need to work with the associated "quantum monoid", which is a certain deformation of the coordinate ring of the algebraic monoid of all $n \times n$ matrices over F. The bialgebra structure of the quantum monoid will allow us to take tensor products of modules for q-Schur algebras in a natural way. We have chosen here to use Manin's quantization of the coordinate ring, see [PW], though one could equally well work with the coordinate ring of [DDo].

Continue with F being a field of characteristic p and assume in addition that $q \in F^\times$ has a square root in F. Let ℓ be the smallest positive integer such that $q^\ell = 1$ (i.e. q is a primitive ℓth root of unity), taking $\ell = 0$ if no such positive integer exists. We fix a square root v of q in F such that if v is a primitive fth root of unity then $f \not\equiv 2 (\mathrm{mod}\, 4)$. Note this is always possible: if ℓ is odd one of $\pm\sqrt{q}$ is again a primitive ℓth root of unity, and if ℓ is even both of $\pm\sqrt{q}$ are primitive 2ℓth roots of unity.

The quantized coordinate ring $A_h = A_{F,v}(h)$ is the associative, unital F-algebra generated by $\{c_{i,j} \mid 1 \le i,j \le h\}$ subject to the relations

$$\begin{aligned}
c_{i,s}c_{j,t} &= c_{j,t}c_{i,s} && (i > j, s < t)\\
c_{i,s}c_{j,t} &= c_{j,t}c_{i,s} + (v - v^{-1})c_{i,t}c_{j,s} && (i > j, s > t)\\
c_{i,s}c_{i,t} &= vc_{i,t}c_{i,s} && (s > t)\\
c_{i,s}c_{j,s} &= vc_{j,s}c_{i,s} && (i > j)
\end{aligned}$$

for all admissible $1 \le i,j,s,t \le h$. Let $I(h,k)$ denote the set of all k-tuples $\underline{i} = (i_1, \dots, i_k)$ of integers between 1 and h as in [G$_2$]. Then, A_h is graded by degree as $A_h = \bigoplus_{k \ge 0} A_{h,k}$, and each homogeneous component $A_{h,k}$ is spanned by *monomials* $c_{\underline{i},\underline{j}} = c_{i_1,j_1}c_{i_2,j_2} \dots c_{i_k,j_k}$ for all 'multi-indices' $\underline{i}, \underline{j} \in I(h,k)$.

There is an F-bialgebra structure on A_h with counit $\varepsilon : A_h \to F$ and comultiplication $\Delta : A_h \to A_h \otimes A_h$ satisfying

$$\varepsilon(c_{\underline{i},\underline{j}}) = \delta_{\underline{i},\underline{j}}, \qquad \Delta(c_{\underline{i},\underline{j}}) = \sum_{\underline{k} \in I(h,k)} c_{\underline{i},\underline{k}} \otimes c_{\underline{k},\underline{j}}$$

for all $\underline{i}, \underline{j} \in I(h,k)$ and $k \geq 0$. The subspace $A_{h,k}$ is a subcoalgebra of A_h. We define the category of all *polynomial representations* (resp. *polynomial representations of degree* k) to be the category comod(A_h) (resp. the category comod$(A_{h,k})$). Note that comod$(A_{h,k})$ is a full subcategory of comod(A_h). Moreover, any A_h-comodule M can be decomposed uniquely as $M = M_0 \oplus \cdots \oplus M_k \oplus \ldots$ where each M_k is polynomial of degree k and $\mathrm{Hom}_{A_h}(M_k, M_l) = 0$ for $k \neq l$.

Let

$$I^2(h,k) = \left\{ (\underline{i}, \underline{j}) \in I(h,k) \times I(h,k) \;\middle|\; \begin{array}{l} j_1 \leq \cdots \leq j_k \text{ and } i_l \leq i_{l+1} \\ \text{whenever } j_l = j_{l+1} \end{array} \right\}.$$

This is a set of representatives of the orbits of Σ_k acting diagonally by place permutation on $I(h,k) \times I(h,k)$. The monomials $\{c_{\underline{i},\underline{j}} \mid (\underline{i}, \underline{j}) \in I^2(h,k)\}$ give a basis for the coalgebra $A_{h,k}$. We let $\{\xi_{\underline{i},\underline{j}} \mid (\underline{i}, \underline{j}) \in I^2(h,k)\}$ denote the corresponding dual basis of the F-linear dual $A^*_{h,k}$. It is known that $A^*_{h,k}$, endowed with the naturally induced algebra structure, is isomorphic to the q-Schur algebra $S_{h,k}$ from §1.2. Moreover, copying the argument of [DDo, 3.2.5] but for the Manin quantization, the isomorphism $A^*_{h,k} \to S_{h,k}$ can be chosen so that $\xi_{\underline{i},\underline{j}}$ corresponds to the natural basis element $\phi^u_{\mu,\lambda}$ from (1.2a) for suitable $\mu, \lambda \vDash k$ and $u \in D_{\mu,\lambda}$, up to multiplying by some power of the unit v. To be precise:

(1.3a) *There is an algebra isomorphism* $A^*_{h,k} \to S_{h,k}$ *under which* $\xi_{\underline{i},\underline{j}}$, *for* $(\underline{i}, \underline{j}) \in I^2(h,k)$, *maps to* $v^{-\ell(u)}\phi^u_{\mu,\lambda}$ *where:*

(i) μ *is the weight of* $\underline{i}$ *(so* $\mu = (m_1, \ldots, m_h)$ *with* m_l *equal to the number of times the integer* l *appears in the tuple* $(i_1, \ldots, i_k)$*);*

(ii) λ *is the weight of* $\underline{j}$*;*

(iii) u *is the unique element of* D^{-1}_μ *such that for each* $l = 1, \ldots, k$, $i_{u^{-1}l}$ *is equal to the* l*th entry of the* k*-tuple* $(\underbrace{1, \ldots, 1}_{m_1 \text{ times}}, \underbrace{2, \ldots, 2}_{m_2 \text{ times}}, \ldots, \underbrace{h, \ldots, h}_{m_h \text{ times}})$ *(*u *is then automatically an element of* $D_{\mu,\lambda}$*);*

Henceforth, we always identify algebras $A^*_{h,k}$ and $S_{h,k}$ according to (1.3a). This allows us to regard a left $S_{h,k}$-module as a right $A_{h,k}$-comodule, and vice versa, in a natural way. In other words, we identify the category mod$(S_{h,k})$ with the full subcategory of comod(A_h) consisting of all A_h-comodules that are finite dimensional and of polynomial degree k. We often switch between these two points of view without comment.

In particular, for $\lambda \in \Lambda^+(h,k)$, we have $A_{h,k}$-comodules, hence A_h-comodules, $L_h(\lambda), \Delta_h(\lambda)$ and $\nabla_h(\lambda)$ induced by the corresponding $S_{h,k}$-modules. The modules

$\{L_h(\lambda) \mid \lambda \in \Lambda^+(h,k)\}$ give a complete set of non-isomorphic irreducibles in the category comod$(A_{h,k})$. It follows directly from this that the modules $\{L_h(\lambda) \mid \lambda \in \Lambda^+(h)\}$ give a complete set of non-isomorphic irreducibles in the category comod(A_h).

The bialgebra structure of A_h endows the category comod(A_h) with a natural notion of tensor product. As a special case, algebra multiplication

$$A_{h,k} \otimes A_{h,l} \to A_{h,k+l}$$

allows us to view the tensor product $M \otimes M'$ of an $A_{h,k}$-comodule M and an $A_{h,l}$-comodule M' as an $A_{h,k+l}$-comodule. Dually, we have an algebra map

$$S_{h,k+l} \to S_{h,k} \otimes S_{h,l}, \tag{1.3.1}$$

which enables us to view the tensor product $M \otimes M'$ of an $S_{h,k}$-module M and an $S_{h,l}$-module M' as an $S_{h,k+l}$-module.

We next introduce notation for various familiar modules. First, let $V_h = L_h((1)) = \Delta_h((1)) = \nabla_h((1))$ be the *natural module*, a right A_h-comodule of dimension h over F. The kth tensor power $V_h \otimes \cdots \otimes V_h$ is naturally a right $A_{h,k}$-comodule, so can be regarded as a left $S_{h,k}$-module. This gives us *tensor space*, which we denote by $V_h^{\otimes k}$. We also have the *symmetric, divided* and *exterior powers*:

$$\begin{aligned} S^k(V_h) &= \nabla_h((k)), \\ Z^k(V_h) &= \Delta_h((k)), \\ \Lambda^k(V_h) &= L_h((1^k)) = \Delta_h((1^k)) = \nabla_h((1^k)). \end{aligned}$$

All these modules actually have more natural direct realizations (which we do not need here) as quotients or submodules of $V_h^{\otimes k}$, see for instance [DDo, §2.1]. More generally, given any composition $\nu = (k_1, \dots, k_a) \models k$, define

$$S^\nu(V_h) = S^{k_1}(V_h) \otimes \cdots \otimes S^{k_a}(V_h), \tag{1.3.2}$$

$$Z^\nu(V_h) = Z^{k_1}(V_h) \otimes \cdots \otimes Z^{k_a}(V_h), \tag{1.3.3}$$

$$\Lambda^\nu(V_h) = \Lambda^{k_1}(V_h) \otimes \cdots \otimes \Lambda^{k_a}(V_h), \tag{1.3.4}$$

all of which can be regarded as right A_h-comodules or left $S_{h,k}$-modules. Observe that in the special case $\nu = (1^k)$ all of $S^\nu(V_h), Z^\nu(V_h)$ and $\Lambda^\nu(V_h)$ are isomorphic simply to $V_h^{\otimes k}$. We will need the following known descriptions of these modules as left ideals of $S_{h,k}$:

(1.3b) *For $\nu \in \Lambda(h,k)$,*

(i) *the left ideal $S_{h,k}\phi^1_{\nu,\nu}$ of $S_{h,k}$ is isomorphic to $Z^\nu(V_h)$ as an $S_{h,k}$-module;*

(ii) *providing $h \geq k$, the left ideal $S_{h,k}\kappa(y_\nu)$ of $S_{h,k}$ is isomorphic to $\Lambda^\nu(V_h)$ as an $S_{h,k}$-module, where $\kappa : H_k \to S_{h,k}$ is the embedding of (1.2b).*

We say an A_h-comodule M has a *Δ-filtration* (resp. a *∇-filtration*) if M has an ascending filtration $0 = M_0 < M_1 < \dots$ with $\sum_i M_i = M$ such that each factor M_i/M_{i-1} is isomorphic to a direct sum of copies of $\Delta_h(\lambda)$ (resp. $\nabla_h(\lambda)$) for some fixed $\lambda \in \Lambda^+(h)$ (depending on i). The following important fact is well-known, see for instance [PW, (10.4.1)]:

(1.3c) *If M, M' are right A_h-comodules having ∇-filtrations (resp. Δ-filtrations) then $M \otimes M'$ also has a ∇-filtration (resp. a Δ-filtration).*

In particular, (1.3c) implies that for any $\nu \in \Lambda(h,k)$, the modules $S^\nu(V_h)$ and $\Lambda^\nu(V_h)$ have ∇-filtrations, while $Z^\nu(V_h)$ and $\Lambda^\nu(V_h)$ have Δ-filtrations. The following fact is well-known; it can be deduced easily from basic properties of modules with Δ- and ∇-filtrations together with the Littlewood-Richardson rule:

(1.3d) *For $\lambda \in \Lambda^+(h,k)$, regard the transpose partition λ' as a composition of k. Then, the space $\mathrm{Hom}_{S_{h,k}}(Z^\lambda(V_h), \Lambda^{\lambda'}(V_h))$ is one dimensional, and the image of any non-zero such homomorphism is isomorphic to $\Delta_h(\lambda)$.*

Suppose now that q is a root of unity (i.e. $\ell > 0$). Then, there is an analogue for A_h (hence for the q-Schur algebras) of Steinberg's tensor product theorem. This was proved in [PW, chapter 9] (for ℓ odd) and [Cl] (in general). Other proofs (for the Dipper-Donkin quantization) have been given in [DDu$_2$, 5.6] or [Do$_7$, §3.2].

Let $\bar{A}_h = A_{F,1}(h)$ denote the free polynomial algebra over F on generators $\{\bar{c}_{i,j} \mid 1 \le i,j \le h\}$ (which is just the above bialgebra A_h in the special case $v = 1$). The comodule representation theory of $\bar{A}_h$ is precisely the classical polynomial representation theory of GL_h as discussed by Green [G$_2$]. For $\lambda \in \Lambda^+(h)$, we will write $\bar{L}_h(\lambda), \bar{\Delta}_h(\lambda)$ and $\bar{\nabla}_h(\lambda)$ for the irreducible, standard and costandard comodules of $\bar{A}_h$, to distinguish them from the ones above for A_h. Of course, these are just the usual polynomial representations of GL_h over F.

Recall that we chose v earlier so that it is a primitive fth root of unity with $f \not\equiv 2 (\mathrm{mod}\, 4)$. So we can apply [PW, chapter 9] for ℓ odd and [Cl] for ℓ even to deduce that there is for each $r \ge 0$ a unique bialgebra homomorphism

$$F_r : \bar{A}_h \to A_h \tag{1.3.5}$$

such that $\bar{c}_{i,j} \mapsto c_{i,j}^{\ell p^r}$ for all $1 \le i,j \le h$. This map is called the *rth Frobenius morphism*. We stress that if $q = 1$ then $\bar{A}_h = A_h, \ell = 1$ and the zeroth Frobenius map F_0 is just the identity map.

Using the Frobenius map F_r, we can regard an $\bar{A}_h$-comodule M with structure map $\tau : M \to M \otimes \bar{A}_h$ as an A_h-comodule with structure map $(\mathrm{id}_M \otimes F_r) \circ \tau : M \to M \otimes A_h$. This gives the *$r$th Frobenius twist* of M denoted $M^{[r]}$. Note that if M is an $\bar{A}_{h,k}$-comodule, then $M^{[r]}$ is an $A_{h,k\ell p^r}$-comodule.

For any $s > 1$, we say that $\lambda = (l_1, \dots, l_h) \in \Lambda^+(h)$ is *s-restricted* if $l_i - l_{i+1} < s$ for $i = 1, \dots, h-1$. By convention, λ is *1-restricted* if and only if $\lambda = (0)$. By an

(ℓ, p)-adic expansion of λ we mean some (non-unique) way of writing

$$\lambda = \lambda_{-1} + \ell\lambda_0 + \ell p\lambda_1 + \ell p^2\lambda_2 + \dots$$

such that $\lambda_{-1} \in \Lambda^+(h)$ is ℓ-restricted and each $\lambda_i \in \Lambda^+(h)$ is p-restricted for $i \geq 0$.

Now we can state the tensor product theorem (after the first twist this is the usual Steinberg tensor product theorem):

(1.3e) *Suppose that $\lambda \in \Lambda^+(h)$ has (ℓ, p)-adic expansion $\lambda = \lambda_{-1} + \ell\lambda_0 + \dots + \ell p^r\lambda_r$. Then, $L_h(\lambda) \cong L_h(\lambda_{-1}) \otimes \bar{L}_h(\lambda_0)^{[0]} \otimes \bar{L}_h(\lambda_1)^{[1]} \cdots \otimes \bar{L}_h(\lambda_r)^{[r]}$.*

The next lemma gives a technical character theoretic fact of importance in §4.3. In the statement, for $\lambda \vdash n$, c_λ is the number of elements of Σ_n of cycle-type λ. We also recall that ϕ_μ denotes the irreducible character of the symmetric group defined in (1.1.2).

1.3f. **Lemma.** *Let ℓ and p be as above. Set $m = \ell p^r$ for some $r \geq 0$ and $k = ml$ for some $l \geq 1$. Then, for any $h \geq k$,*

$$L_h((m^l)) = \frac{(-1)^{k+l}}{l!} \sum_{\lambda \vdash l} \sum_{\mu \vdash k} c_\lambda \phi_\mu(m\lambda) \Delta_h(\mu'),$$

where the equality is written in the Grothendieck group of $\mathrm{mod}(S_{h,k})$.

Proof. This is a calculation involving symmetric functions; we refer the reader to [M, §I.2–I.3] for the basic notions used in the proof. In particular, for $\mu \vdash n$, we will write ε_μ for $(-1)^{n-h(\mu)}$, which is the sign of any element of Σ_n of cycle-type μ.

By (1.3e), the module $L_h((m^l))$ is isomorphic to the Frobenius twist $\bar{L}_h((1^l))^{[r]}$. So, the formal character of $L_h((m^l))$ is the symmetric function

$$\psi = \sum_{1 \leq i_1 < \dots < i_l \leq h} x_{i_1}^m x_{i_2}^m \dots x_{i_l}^m.$$

Now using [M, §I.2, ex.8,10] one sees that ψ can be written as

$$\psi = \frac{1}{l!} \sum_{\lambda \vdash l} \varepsilon_\lambda c_\lambda p_{m\lambda}$$

where $p_{m\lambda}$ is the power sum symmetric function of [M, §I.2]. Now applying [M, I.7.8], we deduce that

$$\psi = \frac{1}{l!} \sum_{\lambda \vdash l} \sum_{\mu \vdash k} \varepsilon_\lambda c_\lambda \phi_\mu(m\lambda) s_\mu$$

where s_μ is the Schur function. Now according to [M, §I.7, ex.2], $\phi_\mu(m\lambda) = \varepsilon_{m\lambda}\phi_{\mu'}(m\lambda)$. Noting that $\varepsilon_\lambda \varepsilon_{m\lambda} = (-1)^{k+l}$, we deduce that

$$\psi = \frac{(-1)^{k+l}}{l!} \sum_{\lambda \vdash l} \sum_{\mu \vdash k} c_\lambda \phi_\mu(m\lambda) s_{\mu'}.$$

Finally, recall by Weyl's character formula that the formal character of $\Delta_h(\mu')$ is precisely the Schur function $s_{\mu'}$. $\square$

Finally in this section, we review the definition of the analogues of Levi subgroups, following [Do7, §4.6]. Fix now $a \geq 1$ and $\mu = (h_1, \dots, h_a) \vDash h$. Let L denote the standard Levi subgroup of $G = GL_h(F)$ consisting of all invertible block diagonal matrices of block sizes $h_1, \dots, h_a$. Let

$$\Omega_\mu = \{(i,j) \in [1,h] \times [1,h] \mid \text{there is some } g \in L \text{ with } g_{ij} \neq 0\}.$$

Define A_μ to be the quotient of A_h by the biideal generated by $\{c_{i,j} \mid 1 \leq i,j \leq h, (i,j) \notin \Omega_\mu\}$. So A_μ is the quantum analogue of the coordinate ring of the monoid corresponding to the Levi subgroup L. We note that

$$A_\mu \cong A_{h_1} \otimes \cdots \otimes A_{h_a} \tag{1.3.6}$$

and that A_μ is graded by degree. So for $k \geq 0$ we can talk about the subcoalgebra $A_{\mu,k}$, which is a coalgebra quotient of $A_{h,k}$; we have the coalgebra isomorphism:

$$A_{\mu,k} \cong \bigoplus_{k_1+\cdots+k_a=k} A_{h_1,k_1} \otimes \cdots \otimes A_{h_a,k_a}. \tag{1.3.7}$$

We have the categories of polynomial and polynomial degree k representations for the Levi subgroup, namely, the categories $\mathrm{comod}(A_\mu)$ and $\mathrm{comod}(A_{\mu,k})$ respectively.

There are Levi analogues of the q-Schur algebra, also discussed by Donkin in [Do7, §4.6]. Define $S_{\mu,k}$ to be the dual space $A^*_{\mu,k}$, with natural algebra structure inherited from the comultiplication and counit of $A_{\mu,k}$. So dualizing (1.3.7), we have that

$$S_{\mu,k} \cong \bigoplus_{k_1+\cdots+k_a=k} S_{h_1,k_1} \otimes \cdots \otimes S_{h_a,k_a}. \tag{1.3.8}$$

We identify $\mathrm{mod}(S_{\mu,k})$ with the full subcategory of $\mathrm{comod}(A_\mu)$ consisting of all finite dimensional A_μ-comodules of polynomial degree k.

Dual to the surjective coalgebra map $A_{h,k} \to A_{\mu,k}$ we have a natural embedding $S_{\mu,k} \hookrightarrow S_{h,k}$. We will always regard $S_{\mu,k}$ as a subalgebra of $S_{h,k}$ embedded in this way. Explicitly, we have that:

(1.3g) *The subalgebra $S_{\mu,k}$ of $S_{h,k}$ is spanned by the standard basis elements $\xi_{\underline{i},\underline{j}}$ for $\underline{i},\underline{j} \in I(h,k)$ such that $(i_l, j_l) \in \Omega_\mu$ for all $l = 1, \dots, k$.*

1.4. Polynomial induction

We recall briefly the definitions of induction and restriction functors for coalgebras, following [Do1]. Let (A, Δ, ε) and $(A', \Delta', \varepsilon')$ be coalgebras over our fixed field F, and $\phi : A \to A'$ be a fixed coalgebra homomorphism. We have the exact *restriction functor*

$$\operatorname{res}^A_{A'} : \operatorname{comod}(A) \to \operatorname{comod}(A')$$

defined on a right A-comodule M with structure map $\tau : M \to M \otimes A$ by letting $\operatorname{res}^A_{A'} M$ be the right A'-comodule equal to M as a vector space but with new structure map $(\operatorname{id}_M \otimes \phi) \circ \tau : M \to M \otimes A'$. On an A-comodule homomorphism $\alpha : M \to M'$, $\operatorname{res}^A_{A'} \alpha$ is the same linear map α, but regarded now as an A'-comodule map. There is a *comodule induction functor* that is right adjoint to $\operatorname{res}^A_{A'}$, namely

$$\operatorname{ind}^A_{A'} : \operatorname{comod}(A') \to \operatorname{comod}(A).$$

To define this on objects, fix $M \in \operatorname{comod}(A')$ with structure map $\tau : M \to M \otimes A'$. Write $|M| \otimes A$ for the right A-comodule which is equal to $M \otimes A$ as a vector space, with structure map $\operatorname{id}_M \otimes \Delta : M \otimes A \to M \otimes A \otimes A$. Then, $\operatorname{ind}^A_{A'} M$ is the subcomodule of $|M| \otimes A$ consisting of all elements f such that

$$(\tau \otimes \operatorname{id}_A)(f) = (\operatorname{id}_M \otimes [(\phi \otimes \operatorname{id}_A) \circ \Delta])(f). \tag{1.4.1}$$

On a morphism $\alpha : M \to M'$, $\operatorname{ind}^A_{A'} \alpha$ is the restriction to $\operatorname{ind}^A_{A'} M$ of the morphism $\alpha \otimes \operatorname{id}_A : |M| \otimes A \to |M'| \otimes A$.

We wish to study induction and restriction (in the sense of coalgebras as just defined) between A_h and its Levi quotient A_μ introduced just before (1.3.6). So choose $\mu \vDash k$ and consider the *polynomial restriction* and *induction* functors:

$$\operatorname{res}^{A_h}_{A_\mu} : \operatorname{comod}(A_h) \to \operatorname{comod}(A_\mu),$$
$$\operatorname{ind}^{A_h}_{A_\mu} : \operatorname{comod}(A_\mu) \to \operatorname{comod}(A_h).$$

We record some basic properties:

1.4a. **Lemma.** (i) *The functor $\operatorname{res}^{A_h}_{A_\mu}$ sends finite dimensional modules to finite dimensional modules, and $A_{h,k}$-comodules to $A_{\mu,k}$-comodules.*

(ii) *The functor $\operatorname{ind}^{A_h}_{A_\mu}$ sends finite dimensional modules to finite dimensional modules, and $A_{\mu,k}$-comodules to $A_{h,k}$-comodules.*

Proof. (i) This is obvious.

(ii) Let M be an $A_{\mu,k}$-comodule with structure map $\tau : M \to M \otimes A_{\mu,k}$. Recall that $A_h = \bigoplus_{l \geq 0} A_{h,l}$ as an A_h-comodule. For $l \geq 0$,

$$\tau \otimes \operatorname{id}_{A_h}(M \otimes A_{h,l}) \subseteq M \otimes A_{\mu,k} \otimes A_{h,l},$$
$$\operatorname{id}_M \otimes [(\phi \otimes \operatorname{id}_{A_h}) \circ \Delta](M \otimes A_{h,l}) \subseteq M \otimes A_{\mu,l} \otimes A_{h,l}.$$

So, recalling the definition of the functor $\operatorname{ind}^{A_h}_{A_\mu}$, we deduce that the only non-zero $f \in M \otimes A_h$ satisfying (1.4.1) in fact lie in $M \otimes A_{h,k}$. Hence, $\operatorname{ind}^{A_h}_{A_\mu}$ sends degree k modules to degree k modules.

Moreover, if M is a finite dimensional $A_{\mu,k}$-module, $\operatorname{ind}_{A_\mu}^{A_h} M$ is finite dimensional because it is a subspace of the finite dimensional space $M \otimes A_{h,k}$. Since any finite dimensional A_μ-comodule can be written as a direct sum of finite dimensional $A_{\mu,k}$-comodules for finitely many different k, we deduce that $\operatorname{ind}_{A_\mu}^{A_h}$ sends finite dimensional modules to finite dimensional modules. □

Now we focus on a special case. So, until just before the end of the section, fix $l \leq h$, set $l' = h - l$ and consider A_μ with $\mu = (l, l') \vDash h$. In this case, A_μ is a bialgebra quotient of A_h isomorphic to $A_l \otimes A_{l'}$; let

$$\phi : A_h \to A_l \otimes A_{l'}$$

be the quotient map. Corresponding to ϕ, we have the coalgebra induction and restriction functors $\operatorname{res}_{A_l \otimes A_{l'}}^{A_h}$ and $\operatorname{ind}_{A_l \otimes A_{l'}}^{A_h}$.

We will also need truncated versions of these functors. There is a natural bialgebra embedding $i : A_l \hookrightarrow A_l \otimes A_{l'}, a \mapsto a \otimes 1$, with image $A_l \otimes A_{l',0}$, recalling that $A_{l',0} \cong F$ denotes the degree zero part of $A_{l'}$. Define the exact truncation functor

$$\operatorname{trunc} : \operatorname{comod}(A_l \otimes A_{l'}) \to \operatorname{comod}(A_l)$$

as follows. On an object M with structure map $\tau : M \to M \otimes A_l \otimes A_{l'}$, $\operatorname{trunc} M$ is defined as the subspace $\{m \in M | \tau(m) \in M \otimes A_l \otimes A_{l',0}\}$, regarded as an A_l-comodule via the restriction of τ and the isomorphism $\operatorname{id}_M \otimes i : M \otimes A_l \to M \otimes A_l \otimes A_{l',0}$. On a morphism, trunc is defined simply as restriction. The functor trunc has an adjoint, namely, the inflation functor

$$\operatorname{infl} : \operatorname{comod}(A_l) \to \operatorname{comod}(A_l \otimes A_{l'})$$

defined on an object M with structure map $\tau : M \to M \otimes A_l$ by letting $\operatorname{infl} M$ be M as a vector space, but with new structure map $\hat{\tau} = (\operatorname{id}_M \otimes i) \circ \tau : M \to M \otimes A_l \otimes A_{l'}$. On a morphism θ, $\operatorname{infl} \theta$ is the same linear map but regarded instead as an $A_l \otimes A_{l'}$-comodule map. Now we define the truncated polynomial restriction and induction functors:

$$\operatorname{trunc}_{A_l}^{A_h} : \operatorname{comod}(A_h) \to \operatorname{comod}(A_l) \quad \text{by} \quad \operatorname{trunc}_{A_l}^{A_h} = \operatorname{trunc} \circ \operatorname{res}_{A_l \otimes A_{l'}}^{A_h},$$
$$\operatorname{infl}_{A_l}^{A_h} : \operatorname{comod}(A_l) \to \operatorname{comod}(A_h) \quad \text{by} \quad \operatorname{infl}_{A_l}^{A_h} = \operatorname{ind}_{A_l \otimes A_{l'}}^{A_h} \circ \operatorname{infl}.$$

Lemma 1.4a easily implies the analogous result for the truncated versions of the functors:

(1.4b) (i) *The functor* $\operatorname{trunc}_{A_l}^{A_h}$ *sends finite dimensional modules to finite dimensional modules, and* $A_{h,k}$*-comodules to* $A_{l,k}$*-comodules.*

(ii) *The functor* $\operatorname{infl}_{A_l}^{A_h}$ *sends finite dimensional modules to finite dimensional modules, and* $A_{l,k}$*-comodules to* $A_{h,k}$*-comodules.*

The main result of the section is as follows:

1.4c. **Theorem.** *For non-negative integers l, l' with $l+l' = h$ as above, the following bifunctors are isomorphic:*

$$\mathrm{ind}^{A_h}_{A_l \otimes A_{l'}}(\,?\boxtimes ?'\,) : \mathrm{comod}(A_l) \times \mathrm{comod}(A_{l'}) \to \mathrm{comod}(A_h),$$
$$(\mathrm{infl}^{A_h}_{A_l}\,?\,) \otimes (\mathrm{infl}^{A_h}_{A_{l'}}\,?'\,) : \mathrm{comod}(A_l) \times \mathrm{comod}(A_{l'}) \to \mathrm{comod}(A_h).$$

Proof. The proof proceeds in a number of steps. We fix an A_l-comodule M and an $A_{l'}$-comodule M'. Let $\tau : M \to M \otimes A_l$ and $\tau' : M' \to M' \otimes A_{l'}$ be their respective structure maps. We denote the structure map of $\mathrm{infl}\, M$ (resp. $\mathrm{infl}\, M'$) as an $A_l \otimes A_{l'}$-comodule by $\hat{\tau} : M \to M \otimes A_l \otimes A_{l'}$ (resp. $\hat{\tau}' : M' \to M' \otimes A_l \otimes A_{l'}$). Define

$$J = \mathrm{span}\{c_{\underline{i},\underline{j}} \mid k \geq 0, \underline{i}, \underline{j} \in I(h,k), 1 \leq i_1, \dots, i_k \leq l\},$$
$$J' = \mathrm{span}\{c_{\underline{i},\underline{j}} \mid k \geq 0, \underline{i}, \underline{j} \in I(h,k), l+1 \leq i_1, \dots, i_k \leq h\}.$$

Note that both J and J' are subalgebras and right A_h-subcomodules of A_h. It is routine to check the following, using the fact that algebra multiplication $\mu : A_h \otimes A_h \to A_h$ is a coalgebra homomorphism:

(1.4d) *Multiplication $\mu : J \otimes J' \to A_h$ is an isomorphism of right A_h-comodules.*

Consider the A_h-comodule maps

$$\theta = (\hat{\tau} \otimes \mathrm{id}_{A_h} - \mathrm{id}_M \otimes [(\phi \otimes \mathrm{id}_{A_h}) \circ \Delta]) : |M| \otimes A_h \to |M \otimes A_l \otimes A_{l'}| \otimes A_h,$$
$$\theta' = (\hat{\tau}' \otimes \mathrm{id}_{A_h} - \mathrm{id}_{M'} \otimes [(\phi \otimes \mathrm{id}_{A_h}) \circ \Delta]) : |M'| \otimes A_h \to |M' \otimes A_l \otimes A_{l'}| \otimes A_h.$$

Note we are using the symbol $|.|$ to emphasize that the A_h-comodule structure is coming just from the final term in these tensor products. We claim that:

(1.4e) $\ker\theta \subseteq |M| \otimes J$ *and* $\ker\theta' \subseteq |M'| \otimes J'$.

We prove (1.4e) just for θ, the proof in the case of θ' being entirely similar. Note directly from the definition of $\hat{\tau}$ that $(\hat{\tau} \otimes \mathrm{id}_{A_h})(\ker\theta) \subseteq M \otimes A_l \otimes A_{l',0} \otimes A_h$. Now take $v \in \ker\theta$ with $v \in M \otimes A_{h,k}$ for some $k \geq 0$. Then, $\bar{v} := (\hat{\tau} \otimes \mathrm{id}_{A_h})(v) = (\mathrm{id}_M \otimes [(\phi \otimes \mathrm{id}_{A_h}) \circ \Delta])(v) \in M \otimes A_{l,k} \otimes A_{l',0} \otimes A_{h,k}$. Write $v = \sum_{\underline{i},\underline{j} \in I(h,k)} m_{\underline{i},\underline{j}} \otimes c_{\underline{i},\underline{j}}$, for some $m_{\underline{i},\underline{j}} \in M$. We have that

$$\bar{v} = \sum_{\underline{i},\underline{j},\underline{s} \in I(h,k)} m_{\underline{i},\underline{j}} \otimes \phi(c_{\underline{i},\underline{s}}) \otimes c_{\underline{s},\underline{j}} \in M \otimes A_{l,k} \otimes A_{l',0} \otimes A_{h,k}. \tag{1.4.2}$$

Consider the projection π_m of $M \otimes \phi(A_{h,k}) \otimes A_{h,k}$ onto the mth term of the direct sum decomposition

$$M \otimes \phi(A_{h,k}) \otimes A_{h,k} = \bigoplus_{m=0}^{k} M \otimes A_{l,k-m} \otimes A_{l',m} \otimes A_{h,k}$$

coming from the grading. We know by (1.4.2) that $\pi_m(\bar{v}) = 0$ for $m > 0$, hence that

$$\bar{v} = \pi_0(\bar{v}) = \sum_{\underline{i},\underline{j},\underline{s} \in I(h,k)} m_{\underline{i},\underline{j}} \otimes \phi(c_{\underline{i},\underline{s}}) \otimes c_{\underline{s},\underline{j}}$$

summing only over $\underline{s}$ with $1 \le s_1, \dots, s_k \le l$. For such $\underline{s}$, $c_{\underline{s},\underline{j}} \in J$, hence $\bar{v} \in M \otimes A_l \otimes A_{l',0} \otimes J$.

The preceding paragraph shows that

$$(\hat{\tau} \otimes \mathrm{id}_{A_h})(\ker \theta) \subseteq M \otimes A_l \otimes A_{l',0} \otimes J. \tag{1.4.3}$$

Consider the map $\mathrm{id}_M \bar{\otimes} \varepsilon : M \otimes A_l \otimes A_{l'} \to M$ such that $m \otimes a \mapsto m\varepsilon(a)$ for $m \in M, a \in A_l \otimes A_{l'}$, where ε is the counit of $A_l \otimes A_{l'}$. By the comodule axioms, $(\mathrm{id}_M \bar{\otimes} \varepsilon) \circ \hat{\tau} = \mathrm{id}_M$. Now applying $\mathrm{id}_M \bar{\otimes} \varepsilon \otimes \mathrm{id}_{A_h}$ to both sides of (1.4.3), we deduce that $\ker \theta \subseteq M \otimes J$ to complete the proof of (1.4e).

Consider now the A_h-comodule map

$$\omega : |M| \otimes J \otimes |M'| \otimes J' \to |M \otimes M'| \otimes A_h$$

defined by $m \otimes j \otimes m' \otimes j' \mapsto m \otimes m' \otimes jj'$. According to (1.4d), this is an isomorphism of A_h-comodules. Observe moreover that $\mathrm{infl}_{A_l}^{A_h} M$ is precisely $\ker \theta$ by the definition (1.4.1), and similarly, $\mathrm{infl}_{A_{l'}}^{A_h} M' = \ker \theta'$. So by (1.4e) and the definitions, we have that

$$(\mathrm{infl}_{A_l}^{A_h} M) \otimes (\mathrm{infl}_{A_{l'}}^{A_h} M') \subseteq |M| \otimes J \otimes |M'| \otimes J',$$
$$\mathrm{ind}_{A_l \otimes A_{l'}}^{A_h} (M \boxtimes M') \subseteq |M \otimes M'| \otimes A_h.$$

We claim that the restriction of ω induces an isomorphism between $(\mathrm{infl}_{A_l}^{A_h} M) \otimes (\mathrm{infl}_{A_{l'}}^{A_h} M')$ and $\mathrm{ind}_{A_l \otimes A_{l'}}^{A_h} (M \boxtimes M')$. This will complete the proof of the theorem, functoriality being immediate as ω is clearly functorial in arguments M and M'.

To prove the claim, ω is a bijection, so we just need to check that

$$\omega^{-1}(\mathrm{ind}_{A_l \otimes A_{l'}}^{A_h} (M \boxtimes M')) = \ker \theta \otimes \ker \theta'. \tag{1.4.4}$$

Recall that $\mathrm{ind}_{A_l \otimes A_{l'}}^{A_h} (M \boxtimes M')$ is the set of vectors in $M \otimes M' \otimes A_h$ satisfying the appropriate version of (1.4.1). So, $\omega^{-1}(\mathrm{ind}_{A_l \otimes A_{l'}}^{A_h} (M \boxtimes M')) = \ker(\alpha - \beta)$, where α and β are the maps from $M \otimes J \otimes M' \otimes J'$ to $M \otimes A_l \otimes M' \otimes A_{l'} \otimes A_h$ defined by

$$\alpha = (\tau \otimes \tau' \otimes \mathrm{id}_{A_h}) \circ \omega \quad \text{and} \quad \beta = (\mathrm{id}_{M \boxtimes M'} \otimes [(\phi \otimes \mathrm{id}_{A_h}) \circ \Delta]) \circ \omega,$$

respectively. Consider instead the maps α' and β' from $M \otimes J \otimes M' \otimes J'$ to $M \otimes A_l \otimes A_{l'} \otimes A_h \otimes M' \otimes A_l \otimes A_{l'} \otimes A_h$ defined by

$$\alpha' = \hat{\tau} \otimes \mathrm{id}_J \otimes \hat{\tau}' \otimes \mathrm{id}_{J'} \quad \text{and} \quad \beta' = \mathrm{id}_M \otimes ([\phi \otimes \mathrm{id}_{A_h}] \circ \Delta) \otimes \mathrm{id}_{M'} \otimes ([\phi \otimes \mathrm{id}_{A_h}] \circ \Delta),$$

respectively. Define

$$\pi : M \otimes A_l \otimes A_{l'} \otimes A_h \otimes M' \otimes A_l \otimes A_{l'} \otimes A_h \to M \otimes A_l \otimes M' \otimes A_{l'} \otimes A_h,$$
$$m \otimes a_1 \otimes a_1' \otimes b_1 \otimes m' \otimes a_2 \otimes a_2' \otimes b_2 \mapsto m \otimes a_1a_2 \otimes m' \otimes a_1'a_2' \otimes b_1b_2.$$

Observe that $\alpha = \pi \circ \alpha'$ and $\beta = \pi \circ \beta'$. Moreover, the images of both α' and β' lie in $Y = M \otimes A_l \otimes A_{l',0} \otimes J \otimes M' \otimes A_{l,0} \otimes A_{l'} \otimes J'$, which is obvious for α' and an easy exercise for β'. Since the restriction of π to Y is injective, it follows that $\ker(\alpha - \beta) = \ker(\alpha' - \beta')$. Writing $\eta = (\alpha' - \beta')$ for short, we have shown that $\omega^{-1}(\mathrm{ind}_{A_l \otimes A_{l'}}^{A_h}(M \boxtimes M')) = \ker \eta$. So (1.4.4) is equivalent to showing:

(1.4f) $\ker \eta = \ker \theta \otimes \ker \theta'$.

To prove (1.4f), first take a pure tensor $k \otimes k' \in \ker \theta \otimes \ker \theta'$. Then, $\hat{\tau} \otimes \mathrm{id}_J(k) = \mathrm{id}_M \otimes ([\phi \otimes \mathrm{id}_{A_h}] \circ \Delta)(k)$ and analogously for k', which immediately shows that $k \otimes k' \in \ker \eta$. Conversely, let

$$\mathrm{id}_M \bar{\otimes} \varepsilon \bar{\otimes} \mathrm{id}_J : M \otimes A_l \otimes A_{l'} \otimes J \to M \otimes J$$

be the map $m \otimes c \otimes j \mapsto m\varepsilon(c) \otimes j = m \otimes \varepsilon(c)j$, where ε is the counit of $A_l \otimes A_{l'}$. Recalling that ϕ is a bialgebra map, the coalgebra and comodule axioms immediately give that both of the composites

$$(\mathrm{id}_M \bar{\otimes} \varepsilon \bar{\otimes} \mathrm{id}_J) \circ (\hat{\tau} \otimes \mathrm{id}_J) : M \otimes J \to M \otimes J,$$
$$(\mathrm{id}_M \bar{\otimes} \varepsilon \bar{\otimes} \mathrm{id}_J) \circ (\mathrm{id}_M \otimes [(\phi \otimes \mathrm{id}_{A_h}) \circ \Delta]) : M \otimes J \to M \otimes J$$

are equal to the identity. So,

$$(\mathrm{id}_M \bar{\otimes} \varepsilon \bar{\otimes} \mathrm{id}_J \otimes \mathrm{id}_{M' \otimes A_l \otimes A_{l'} \otimes J'}) \circ \eta = \mathrm{id}_{M \otimes J} \otimes \theta'.$$

It follows directly that $\ker \eta \subseteq \ker(\mathrm{id}_{M \otimes J} \otimes \theta') = M \otimes J \otimes \ker \theta'$. A similar argument shows that $\ker \eta \subseteq \ker \theta \otimes M' \otimes J'$. Hence,

$$\ker \eta \subseteq (\ker \theta \otimes M' \otimes J') \cap (M \otimes J \otimes \ker \theta'). \tag{1.4.5}$$

Finally, by linear algebra the right hand side of (1.4.5) is precisely $\ker \theta \otimes \ker \theta'$, which completes the proof of (1.4f) hence the theorem. □

Finally, we extend Theorem 1.4c to the general case, to obtain the quantum analogue of [BK$_1$, Theorem 2.7]:

1.4g. **Corollary.** *Fix $\mu = (h_1, \dots, h_a) \models h$. Then, the following functors are isomorphic:*

$$\mathrm{ind}_{A_\mu}^{A_h}(\ ?\ \boxtimes \cdots \boxtimes\ ?\ ) : \mathrm{comod}(A_{h_1}) \times \cdots \times \mathrm{comod}(A_{h_a}) \to \mathrm{comod}(A_h),$$
$$(\mathrm{infl}_{A_{h_1}}^{A_h}\ ?\ ) \otimes \cdots \otimes (\mathrm{infl}_{A_{h_a}}^{A_h}\ ?\ ) : \mathrm{comod}(A_{h_1}) \times \cdots \times \mathrm{comod}(A_{h_a}) \to \mathrm{comod}(A_h).$$

Proof. This follows from Theorem 1.4c by induction on a, using the fact that coalgebra induction is transitive. □

We refer the reader to [BK$_1$, §2] for further properties of the polynomial induction and restriction functors, as well as some consequences of Corollary 1.4g, in the classical case. The proofs in *loc. cit.* carry over to the quantum case, now that we have Corollary 1.4g.

1.5. Schur algebra induction

The goal in this section is to reformulate Corollary 1.4g in terms of the q-Schur algebra. Fix throughout the section $h, k \geq 1$ and $\mu = (h_1, \dots, h_a) \vDash h$. Recall that $S_{\mu,k}$ denotes the Levi subalgebra of $S_{h,k}$, defined as in (1.3.8), over the field F. We have the restriction and induction functors

$$\begin{aligned} \operatorname{res}^{S_{h,k}}_{S_{\mu,k}} &: \operatorname{mod}(S_{h,k}) \to \operatorname{mod}(S_{\mu,k}), \\ \operatorname{ind}^{S_{h,k}}_{S_{\mu,k}} &: \operatorname{mod}(S_{\mu,k}) \to \operatorname{mod}(S_{h,k}) \end{aligned}$$

in the usual sense of finite dimensional algebras; so $\operatorname{ind}^{S_{h,k}}_{S_{\mu,k}} = S_{h,k} \otimes_{S_{\mu,k}} ?$. Note that $\operatorname{ind}^{S_{h,k}}_{S_{\mu,k}}$ is left adjoint to $\operatorname{res}^{S_{h,k}}_{S_{\mu,k}}$.

We also need truncated versions, so fix $l \leq h$ and embed $\Lambda(l,k)$ (resp. $I(l,k)$) in $\Lambda(h,k)$ (resp. $I(h,k)$) in the natural way. Let e be the idempotent

$$e = e_{h,l} = \sum_{\mu \in \Lambda(l,k)} \phi^1_{\mu,\mu} \in S_{h,k}. \tag{1.5.1}$$

Note this is the analogue of the idempotent defined by Green in [G$_2$, (6.5b)]. The subring $eS_{h,k}e$ is spanned by all standard basis elements $\xi_{\underline{i},\underline{j}}$ with $\underline{i}, \underline{j} \in I(l,k)$. Moreover, just as in the classical case [G$_2$, p. 103],

$$S_{l,k} \cong eS_{h,k}e,$$

the natural basis element $\xi_{\underline{i},\underline{j}}$ of $S_{l,k}$ for $\underline{i}, \underline{j} \in I(l,k)$ mapping under the isomorphism to the corresponding natural basis element $\xi_{\underline{i},\underline{j}}$ of $S_{h,k}$. In what follows, we identify $S_{l,k}$ with $eS_{h,k}e$ in this way. Then, we have the Schur functor

$$\operatorname{trunc}^{S_{h,k}}_{S_{l,k}} : \operatorname{mod}(S_{h,k}) \to \operatorname{mod}(S_{l,k}), \tag{1.5.2}$$

defined on an object M by $\operatorname{trunc}^{S_{h,k}}_{S_{l,k}} M = eM$ and by restriction on morphisms. The functor $\operatorname{trunc}^{S_{h,k}}_{S_{l,k}}$ has a left adjoint, the inverse Schur functor

$$\operatorname{infl}^{S_{h,k}}_{S_{l,k}} : \operatorname{mod}(S_{l,k}) \to \operatorname{mod}(S_{h,k}), \tag{1.5.3}$$

which is the functor $S_{h,k}e \otimes_{eS_{h,k}e} ?$.

The truncation functor $\operatorname{trunc}^{S_{h,k}}_{S_{l,k}}$ is the quantum analogue of the functor $d_{h,l}$ considered by Green [G$_2$, §6.5], and the argument of [G$_2$, (6.5g)] carries over to show:

(1.5a) *Assuming that $k \leq l$, the functors*

$$\mathrm{trunc}_{S_{l,k}}^{S_{h,k}} : \mathrm{mod}(S_{h,k}) \to \mathrm{mod}(S_{l,k}) \qquad \textit{and} \qquad \mathrm{infl}_{S_{l,k}}^{S_{h,k}} : \mathrm{mod}(S_{l,k}) \to \mathrm{mod}(S_{h,k})$$

are mutually inverse equivalences of categories.

The effect of the truncation functor $\mathrm{trunc}_{S_{l,k}}^{S_{h,k}}$ on the standard $S_{h,k}$-modules is well-known; we record the basic facts, see e.g. [G$_2$, (6.5f)], [BK$_1$, 2.3, 2.4], [Do$_7$, §4.2]:

(1.5b) *Take $\mu = (m_1, \dots, m_h) \in \Lambda^+(h,k)$.*
(i) *If $m_{l+1} \neq 0$ then $\mathrm{trunc}_{S_{l,k}}^{S_{h,k}} L_h(\mu) = \mathrm{trunc}_{S_{l,k}}^{S_{h,k}} \Delta_h(\mu) = \mathrm{trunc}_{S_{l,k}}^{S_{h,k}} \nabla_h(\mu) = 0$.*
(ii) *If $m_{l+1} = 0$, we may regard μ as an element of $\Lambda^+(l,k) \subseteq \Lambda^+(h,k)$, and then $\mathrm{trunc}_{S_{l,k}}^{S_{h,k}} L_h(\mu) \cong L_l(\mu)$, $\mathrm{trunc}_{S_{l,k}}^{S_{h,k}} \Delta_h(\mu) \cong \Delta_l(\mu)$ and $\mathrm{trunc}_{S_{l,k}}^{S_{h,k}} \nabla_h(\mu) \cong \nabla_l(\mu)$.*

Now we wish to relate these Schur algebra functors to the polynomial induction and restriction functors of the previous section.

First, recall from §1.2 that the algebra $S_{h,k}$ has an anti-automorphism τ, which is easily seen to stabilize the subalgebra $S_{\mu,k}$, for instance using (1.3g). This allows us to define the contravariant dual M^τ of an $S_{\mu,k}$-module M, in the same way as we did for $S_{h,k}$ in §1.2. Contravariant duality gives us functors $\mathrm{mod}(S_{h,k}) \to \mathrm{mod}(S_{h,k})$, $\mathrm{mod}(S_{\mu,k}) \to \mathrm{mod}(S_{\mu,k})$ and $\mathrm{mod}(S_{l,k}) \to \mathrm{mod}(S_{l,k})$, all of which we will denote simply by τ. It is obvious that τ commutes with $\mathrm{res}_{S_{\mu,k}}^{S_{h,k}}$ and $\mathrm{trunc}_{S_{l,k}}^{S_{h,k}}$. To be precise, there are isomorphisms of functors:

$$\tau \circ \mathrm{res}_{S_{\mu,k}}^{S_{h,k}} \cong \mathrm{res}_{S_{\mu,k}}^{S_{h,k}} \circ \tau, \tag{1.5.4}$$

$$\tau \circ \mathrm{trunc}_{S_{l,k}}^{S_{h,k}} \cong \mathrm{trunc}_{S_{l,k}}^{S_{h,k}} \circ \tau \tag{1.5.5}$$

We remark that contravariant duality does not in general commute with the functor $\mathrm{ind}_{S_{\mu,k}}^{S_{h,k}}$ (resp. $\mathrm{infl}_{S_{l,k}}^{S_{h,k}}$ unless $k \leq l$).

Second, notice that, in view of Lemma 1.4a and (1.4b), we can restrict the functors $\mathrm{res}_{A_\mu}^{A_h}$, $\mathrm{ind}_{A_\mu}^{A_h}$, $\mathrm{res}_{A_l}^{A_h}$ and $\mathrm{ind}_{A_l}^{A_h}$ to finite dimensional, polynomial degree k modules to obtain well-defined functors which we will denote by the same names:

$$\begin{aligned} \mathrm{res}_{A_\mu}^{A_h} &: \mathrm{mod}(S_{h,k}) \to \mathrm{mod}(S_{\mu,k}), \\ \mathrm{ind}_{A_\mu}^{A_h} &: \mathrm{mod}(S_{\mu,k}) \to \mathrm{mod}(S_{h,k}), \\ \mathrm{trunc}_{A_l}^{A_h} &: \mathrm{mod}(S_{h,k}) \to \mathrm{mod}(S_{l,k}), \\ \mathrm{infl}_{A_l}^{A_h} &: \mathrm{mod}(S_{l,k}) \to \mathrm{mod}(S_{h,k}). \end{aligned}$$

To connect all these functors, we have:

1.5c. **Lemma.** *The following pairs of functors are isomorphic:*
(i) $\mathrm{res}^{S_{h,k}}_{S_{\mu,k}}$ *and* $\mathrm{res}^{A_h}_{A_\mu} : \mathrm{mod}(S_{h,k}) \to \mathrm{mod}(S_{\mu,k})$;
(ii) $\mathrm{trunc}^{S_{h,k}}_{S_{l,k}}$ *and* $\mathrm{trunc}^{A_h}_{A_l} : \mathrm{mod}(S_{h,k}) \to \mathrm{mod}(S_{l,k})$;
(iii) $\tau \circ \mathrm{ind}^{S_{h,k}}_{S_{\mu,k}} \circ \tau$ *and* $\mathrm{ind}^{A_h}_{A_\mu} : \mathrm{mod}(S_{\mu,k}) \to \mathrm{mod}(S_{h,k})$;
(iv) $\tau \circ \mathrm{infl}^{S_{h,k}}_{S_{l,k}} \circ \tau$ *and* $\mathrm{infl}^{A_h}_{A_l} : \mathrm{mod}(S_{l,k}) \to \mathrm{mod}(S_{h,k})$.

Proof. (i) This is obvious.

(ii) To see this, note that both of the functors $\mathrm{trunc}^{S_{h,k}}_{S_{l,k}}$ and $\mathrm{trunc}^{A_h}_{A_l}$ are defined on objects by taking certain weight spaces, and by restriction on morphisms. To be more precise, on an object M, both functors send M to $\bigoplus_{\lambda \in \Lambda(l,k)} \phi^1_{\lambda,\lambda} M$. Hence, the functors are isomorphic.

(iii) Using (i) and the fact that $\mathrm{ind}^{A_h}_{A_\mu}$ is right adjoint to $\mathrm{res}^{A_h}_{A_\mu}$, we just need to check by uniqueness of adjoint functors that $\tau \circ \mathrm{ind}^{S_{h,k}}_{S_{\mu,k}} \circ \tau$ is right adjoint to $\mathrm{res}^{S_{h,k}}_{S_{\mu,k}}$. Well, for $M \in \mathrm{mod}(S_{h,k}), N \in \mathrm{mod}(S_{\mu,k})$, we have

$$\begin{aligned}\mathrm{Hom}_{S_{h,k}}(M, (\mathrm{ind}^{S_{h,k}}_{S_{\mu,k}} N^\tau)^\tau) &\cong \mathrm{Hom}_{S_{h,k}}(\mathrm{ind}^{S_{h,k}}_{S_{\mu,k}} N^\tau, M^\tau) \\ &\cong \mathrm{Hom}_{S_{\mu,k}}(N^\tau, \mathrm{res}^{S_{h,k}}_{S_{\mu,k}}(M^\tau)) \\ &\cong \mathrm{Hom}_{S_{\mu,k}}(\mathrm{res}^{S_{h,k}}_{S_{\mu,k}} M, N)\end{aligned}$$

using the fact that contravariant duality commutes with restriction (1.5.4).

(iv) This follows by (ii) and (1.5.5) by the same general argument as (iii). □

Finally, we can restate Corollary 1.4g in terms of Schur algebras. To explain the notation in the statement, note that given modules $M_1, \dots, M_a$ with each $M_i \in \mathrm{mod}(S_{h_i,k_i})$, we can regard the outer tensor product $M_1 \boxtimes \cdots \boxtimes M_a$ as an $S_{\mu,k}$-module where $k = k_1 + \cdots + k_a$. We obtain in this way a functor

$$? \boxtimes \cdots \boxtimes ? : \mathrm{mod}(S_{h_1,k_1}) \times \cdots \times \mathrm{mod}(S_{h_a,k_a}) \to \mathrm{mod}(S_{\mu,k}).$$

Similarly, the tensor product operation on modules over Schur algebras, induced via the map (1.3.1), gives us a functor

$$? \otimes \cdots \otimes ? : \mathrm{mod}(S_{h,k_1}) \times \cdots \times \mathrm{mod}(S_{h,k_a}) \to \mathrm{mod}(S_{h,k})$$

where $k = k_1 + \cdots + k_a$. With this notation:

1.5d. **Theorem.** *For* $\mu = (h_1, \dots, h_a) \vDash h$ *and* $\nu = (k_1, \dots, k_a) \vDash k$, *the following functors are isomorphic:*

$$\mathrm{ind}^{S_{h,k}}_{S_{\mu,k}}(? \boxtimes \cdots \boxtimes ?) : \mathrm{mod}(S_{h_1,k_1}) \times \cdots \times \mathrm{mod}(S_{h_a,k_a}) \to \mathrm{mod}(S_{h,k}),$$

$$(\mathrm{infl}^{S_{h,k_1}}_{S_{h_1,k_1}} ?) \otimes \cdots \otimes (\mathrm{infl}^{S_{h,k_a}}_{S_{h_a,k_a}} ?) : \mathrm{mod}(S_{h_1,k_1}) \times \cdots \times \mathrm{mod}(S_{h_a,k_a}) \to \mathrm{mod}(S_{h,k}).$$

Proof. Combine Corollary 1.4g and Lemma 1.5c(iii),(iv). □

Later, Theorem 1.5d will be important because it gives a way to realize the tensor product operation induced by the bialgebra structure of A_h directly within the finite dimensional algebras $S_{h,k}$ (for all k).

Chapter 2

Classical results on GL_n

We next collect all the results about $GL_n(\mathbb{F}_q)$ from the literature that we will need later. Most of these results are of a purely character theoretic nature. At the same time, we will deduce their basic consequences for the modular theory using elementary base change arguments.

2.1. Conjugacy classes and Levi subgroups

We begin with some basic notation that will be in place for the remainder of the article. Choose some prime power q and denote the finite general linear group $GL_n(\mathbb{F}_q)$ by G_n. Let p be a prime *not dividing* q, F be an algebraically closed field of characteristic p and fix a p-modular system $(F, \mathcal{O}, K)$ with K sufficiently large (see [Ka, §3.1]). So $\mathcal{O}$ is a complete discrete valuation ring with residue field F and field of fractions K of characteristic 0, and moreover K is a splitting field for all finite groups that we meet.

For $\sigma \in \bar{\mathbb{F}}_q^\times$, we let (σ) denote the associated *companion matrix*. So if σ is of degree d over $\mathbb{F}_q$, (σ) is the $d \times d$-matrix corresponding to the automorphism of $\mathbb{F}_q[\sigma]$ induced by left multiplication by σ, when written in terms of the $\mathbb{F}_q$-basis $1, \sigma, \dots, \sigma^{d-1}$. We say that σ is *p-regular* if its multiplicative order is coprime to p; otherwise, we say that σ is *p-singular*. Given in addition $\tau \in \bar{\mathbb{F}}_q^\times$, we say that σ and τ are *conjugate* if they have the same minimal polynomial over $\mathbb{F}_q$, or equivalently, if the matrices (σ) and (τ) are conjugate matrices.

For $d \geq 1$, let $\ell(d)$ denote the smallest positive integer such that $q^{d\ell(d)} \equiv 1 (\operatorname{mod} p)$ (i.e. the image of q^d in F is a primitive $\ell(d)$th root of unity). We record the basic number theoretic fact (cf. [DJ$_2$, Lemma 2.3]):

(2.1a) *Let $\sigma \in \bar{\mathbb{F}}_q^\times$ be a p-regular element of degree d over $\mathbb{F}_q$. There exists a p-singular element $\tau \in \bar{\mathbb{F}}_q^\times$ of degree e over $\mathbb{F}_q$ with p-regular part conjugate to σ if and only if $e = d\ell(d)p^r$ for some $r \geq 0$.*

For $\sigma \in \bar{\mathbb{F}}_q^\times$ and $k \geq 1$, let $(\sigma)^k$ denote the block diagonal matrix consisting of k copies of (σ) along the diagonal. Every semisimple conjugacy class of G_n can be represented as a block diagonal matrix

$$s = \mathrm{diag}((\sigma_1)^{k_1}, \dots, (\sigma_a)^{k_a}) \tag{2.1.1}$$

for $\sigma_i \in \bar{\mathbb{F}}_q^\times$ with σ_i not conjugate to σ_j for $i \neq j$, and integers $k_1, \dots, k_a \geq 0$. We will say $s \in G_n$ is *block-diagonal* if it is a semisimple element of the form (2.1.1). For such an s, we associate the following compositions:

$$\kappa(s) = (k_1, \dots, k_a), \tag{2.1.2}$$
$$\delta(s) = ((d_1)^{k_1}, \dots, (d_a)^{k_a}) \models n, \tag{2.1.3}$$
$$\pi(s) = (d_1 k_1, \dots, d_a k_a) \models n, \tag{2.1.4}$$

where d_i is the degree of σ_i over $\mathbb{F}_q$. The centralizer $C_{G_n}(s)$ is isomorphic to

$$GL_{k_1}(\mathbb{F}_{q^{d_1}}) \times \cdots \times GL_{k_a}(\mathbb{F}_{q^{d_a}}). \tag{2.1.5}$$

Recalling that the unipotent classes of G_n are parametrized in a standard way by partitions $\lambda \vdash n$, we see from (2.1.5) that the unipotent classes of $C_{G_n}(s)$ are parametrized by *multi-partitions* $\underline{\lambda} = (\lambda_1, \dots, \lambda_a)$ where $\lambda_i \vdash k_i$ for $i = 1, \dots, a$. We will write $\underline{\lambda} \vdash \kappa(s)$ if $\underline{\lambda}$ is such a multi-partition.

Let $\mathcal{C}_{ss}$ be a set of representatives of the semisimple classes of G_n, such that each $s \in \mathcal{C}_{ss}$ is block-diagonal, and let $\mathcal{C}_{ss,p'}$ denote the p-regular elements $s \in \mathcal{C}_{ss}$, that is, the elements of $\mathcal{C}_{ss}$ of order coprime to p. By the Jordan decomposition, the conjugacy classes of G_n are parametrized by pairs $(s, \underline{\lambda})$ for all $s \in \mathcal{C}_{ss}$ and $\underline{\lambda} \vdash \kappa(s)$. Moreover, as p is coprime to q, the p-regular conjugacy classes are parametrized by pairs $(s, \underline{\lambda})$ for all $s \in \mathcal{C}_{ss,p'}$ and $\underline{\lambda} \vdash \kappa(s)$.

We turn next to describing various Levi subgroups of G_n. Let $\underline{G}_n = GL_n(\bar{\mathbb{F}}_q)$ denote the corresponding algebraic group and

$$f_q : \underline{G}_n \to \underline{G}_n \tag{2.1.6}$$

be the Frobenius map defined by raising all the entries of a matrix $g \in \underline{G}_n$ to the qth power. Then, the finite general linear group G_n is precisely the set of f_q-fixed points in $\underline{G}_n$. By a *Levi subgroup* of G_n, we mean the set of f_q-fixed points in an f_q-stable Levi subgroup of $\underline{G}_n$. For example, the centralizer $C_{G_n}(s)$ of (2.1.5) is a Levi subgroup of G since it is the set of f_q-fixed points in the Levi subgroup $C_{\underline{G}_n}(s)$ of $\underline{G}_n$.

Another special case gives us maximal tori: a *maximal torus* of G_n is the set of f_q-fixed points in an f_q-stable maximal torus of $\underline{G}_n$. By Lang's theorem, the conjugacy classes of maximal tori of G_n are parametrized by conjugacy classes of the symmetric group Σ_n, hence by partitions of n. For $\lambda \vdash n$, we let T_λ be a representative of the corresponding conjugacy class of maximal tori of G_n, so

$$T_\lambda \cong GL_1(\mathbb{F}_{q^{l_1}}) \times \cdots \times GL_1(\mathbb{F}_{q^{l_a}}) \tag{2.1.7}$$

if $\lambda = (l_1, \dots, l_a) \vdash n$.

Let $G_\nu = G_{n_1} \times \cdots \times G_{n_a}$ (embedded into G_n as block diagonal matrices) denote the *standard Levi subgroup* of G_n parametrized by the composition $\nu = (n_1, \ldots, n_a) \models n$. We write U_n (resp., U_ν) for the subgroup of all upper uni-triangular matrices in G_n (resp., G_ν) and Y_ν for the unipotent radical of the standard parabolic subgroup of G_n with Levi factor G_ν, so $U_n = U_\nu Y_\nu$ (semi-direct product).

We always henceforth identify the symmetric group Σ_n with the subgroup of G_n consisting of all permutation matrices, so that for $\nu \models n$, $\Sigma_\nu = \Sigma_n \cap G_\nu$. The following fact follows from (1.1a):

(2.1b) *Given $\lambda, \mu \models n$, where again $n = kd$, and $w \in D_{\mu,\lambda}$, $G_\mu \cap {}^w G_\lambda$ is a standard Levi subgroup of G_n.*

Now suppose that $n = kd$ for integers $k, d \geq 1$. There is an embedding of Σ_k into G_n as the subgroup of all "$d \times d$-block permutation matrices", so for example, the basic transposition $(1\ 2) \in \Sigma_3$ corresponds to the matrix

$$\begin{pmatrix} 0 & I_d & 0 \\ I_d & 0 & 0 \\ 0 & 0 & I_d \end{pmatrix}$$

in G_{3d}. For $x \in \Sigma_k$, let π_x denote the corresponding block permutation matrix in G_n. If $\nu = (k_1, \ldots, k_a) \models k$, we write $d\nu$ for the composition $(dk_1, \ldots, dk_a) \models n$. We will often appeal to the following observation. Although the proof is not immediate, it is a purely combinatorial statement about coset representatives in the symmetric group which we leave as an exercise for the reader, referring to [DJ$_1$, p. 23] for guidance.

(2.1c) *Given $\lambda, \mu \models k$ and $w \in D_{d\mu,d\lambda}$, $G_{d\mu} \cap {}^w G_{d\lambda}$ contains $G_{(d^k)}$ if and only if $w = \pi_x$ for some $x \in D_{\mu,\lambda}$. In that case, $G_{d\mu} \cap {}^w G_{d\lambda} = G_{d\nu}$ where $\nu \models k$ is determined by $G_\mu \cap {}^x G_\lambda = G_\nu$.*

2.2. Harish-Chandra induction and restriction

Let $G = G_\nu$ for some $\nu \models n$ and L be any standard Levi subgroup of G. Let P be the standard parabolic subgroup of G with Levi factor L. Let Y denote the unipotent radical of P, so that there is a surjection $P \to L$ with kernel Y. We write infl_L^P for the usual inflation functor from $\mathrm{mod}(FL)$ to $\mathrm{mod}(FP)$ along this surjection, and inv_Y for the truncation functor from $\mathrm{mod}(FP)$ to $\mathrm{mod}(FL)$ induced by taking Y-fixed points. Define the *Harish-Chandra induction* and *restriction functors*

$$R_L^G : \mathrm{mod}(FL) \to \mathrm{mod}(FG) \quad \text{by} \quad R_L^G = \mathrm{ind}_P^G \circ \mathrm{infl}_L^P,$$
$${}^*R_L^G : \mathrm{mod}(FG) \to \mathrm{mod}(FL) \quad \text{by} \quad {}^*R_L^G = \mathrm{inv}_Y \circ \mathrm{res}_P^G.$$

It is known (cf. [DDu$_1$, §5], [HL$_2$]) that if some other parabolic subgroup is used in these definitions, one obtains isomorphic functors. However, it is then not immediately obvious that our later definitions (specifically, the modules defined in §3.3) are independent of this choice. This is indeed the case, as shown in [Ac], but for simplicity, we almost always work with the fixed choice of the standard parabolic in this article.

We record two well-known basic properties of the functors R_L^G and ${}^*R_L^G$; the first depends on the assumption that $(p,q)=1$. See [DF, §1] for proofs.

(2.2a) *The functors* ${}^*R_L^G$ *and* R_L^G *are both exact, and are both left and right adjoint to one another.*

(2.2b) *Given standard Levi subgroups* $L' \leq L \leq G$, *the functors* $R_L^G \circ R_{L'}^L$ *and* $R_{L'}^G$ *(resp.,* ${}^*R_{L'}^L \circ {}^*R_L^G$ *and* ${}^*R_{L'}^G$*) are isomorphic.*

We will need the fundamental Mackey decomposition theorem (see e.g. [DF, Theorem 1.14]), which makes sense in view of (2.1b):

(2.2c) *Given standard Levi subgroups* G_μ, G_λ *of* $G = G_\nu$ *for* $\lambda, \mu \vDash n$, *there is an isomorphism of functors*

$$ {}^*R^G_{G_\mu} \circ R^G_{G_\lambda} \cong \bigoplus_{w \in D^\nu_{\mu,\lambda}} R^{G_\mu}_{G_\mu \cap {}^w G_\lambda} \circ \mathrm{conj}_w \circ {}^*R^{G_\lambda}_{G_\lambda \cap {}^{w^{-1}} G_\mu}, $$

where $\mathrm{conj}_w : \mathrm{mod}(F(G_\lambda \cap {}^{w^{-1}} G_\mu)) \to \mathrm{mod}(F(G_\mu \cap {}^w G_\lambda))$ *denotes the functor induced by conjugation by* $w \in G$.

The functors R_L^G and ${}^*R_L^G$ can also be defined in the same way as above over the ground ring $\mathcal{O}$ (similarly, over K). This gives us functors which we will denote with the same names, namely, $R_L^G : \mathrm{mod}(\mathcal{O}L) \to \mathrm{mod}(\mathcal{O}G)$ and ${}^*R_L^G : \mathrm{mod}(\mathcal{O}G) \to \mathrm{mod}(\mathcal{O}L)$. The functors over $\mathcal{O}$ commute with base change, so this notation should not cause confusion:

2.2d. **Lemma.** (i) *For any* $\mathcal{O}L$*-module* $M_\mathcal{O}$, *the* FG*-modules* $F \otimes_\mathcal{O} (R_L^G M_\mathcal{O})$ *and* $R_L^G(F \otimes_\mathcal{O} M_\mathcal{O})$ *are naturally isomorphic.*

(ii) *For any* $\mathcal{O}G$*-module* $N_\mathcal{O}$, *the* FL*-modules* $F \otimes_\mathcal{O} ({}^*R_L^G N_\mathcal{O})$ *and* ${}^*R_L^G(F \otimes_\mathcal{O} N_\mathcal{O})$ *are naturally isomorphic.*

Proof. These results are well-known, but since we could not find a suitable reference we include a proof. First, let H be any group and A (resp. B) be a right (resp. left) $\mathcal{O}H$-module. Write $A_F = F \otimes_\mathcal{O} A$ and $B_F = F \otimes_\mathcal{O} B$. Then

$$ A_F \cong A \otimes_\mathcal{O} F \cong (A \otimes_{\mathcal{O}H} \mathcal{O}H) \otimes_\mathcal{O} F \cong A \otimes_{\mathcal{O}H} (\mathcal{O}H \otimes_\mathcal{O} F) = A \otimes_{\mathcal{O}H} FH. $$

Similarly, $B_F \cong FH \otimes_{\mathcal{O}H} B$. Now

$$ \begin{aligned} A_F \otimes_{FH} B_F &\cong (A \otimes_{\mathcal{O}H} FH) \otimes_{FH} (FH \otimes_{\mathcal{O}H} B) \cong (A \otimes_{\mathcal{O}H} FH) \otimes_{\mathcal{O}H} B \\ &\cong (A \otimes_\mathcal{O} F) \otimes_{\mathcal{O}H} B \cong (F \otimes_\mathcal{O} A) \otimes_{\mathcal{O}H} B \\ &\cong F \otimes_\mathcal{O} (A \otimes_{\mathcal{O}H} B). \end{aligned} $$

Now take $H = L, B = M_{\mathcal{O}}$ and $A = \mathcal{O}Ge$ where e is the idempotent $\frac{1}{|Y|}\sum_{y\in Y} y$. Then, $R_L^G M_{\mathcal{O}} = A \otimes_{\mathcal{O}H} B$, while $R_L^G M_F = A_F \otimes_{FH} B_F$. So (i) follows directly from the general fact just proved. The proof of (ii) is entirely similar, since Harish-Chandra restriction can also be interpreted as tensoring with a certain bimodule. □

We also need to know that the basic properties of Harish-Chandra induction and restriction, namely, (2.2a), (2.2b) and (2.2c), are true also over $\mathcal{O}$; indeed, the references cited above prove the results over any ground ring in which q is invertible.

In the category mod(FG), there is a notion of *contravariant duality*. Let $\tau : G \to G$ denote the anti-automorphism given by matrix transposition. Given a left FG-module V let V^τ denote the left FG-module which as an F-space is equal to the dual $V^* = \mathrm{Hom}_F(V, F)$, but with action defined by $(g.f)(v) = f(\tau(g)v)$ for $v \in V, g \in G, f \in V^*$. Since τ leaves conjugacy classes of G invariant, V and V^τ have the same Brauer character. In particular, if V is an irreducible FG-module, then $V \cong V^\tau$ (cf. [J$_2$, (7.27)]). The same remarks apply to the standard Levi subgroup L of G, since τ stabilizes any such subgroup.

2.2e. **Lemma.** *For a left FL-module V, $R_L^G(V^\tau) \cong (R_L^G V)^\tau$. Similarly, for a left FG-module U, ${}^*R_L^G(U^\tau) \cong ({}^*R_L^G U)^\tau$.*

Proof. We first prove that the functors $(?)^\tau \circ {}^*R_L^G \circ (?)^\tau$ and ${}^*R_L^G$ are isomorphic. Let P be the standard parabolic subgroup of G with Levi factor L and unipotent radical Y, and let $e = \frac{1}{|Y|}\sum_{u\in Y} u \in FG$. Then, the functor ${}^*R_L^G$ is given on objects by left multiplication by the idempotent e, and by restriction on morphisms. The corresponding idempotent for the opposite parabolic subgroup $\tau(P)$ of G containing L is $\tau(e)$, and by [DDu$_1$, §5] or [HL$_2$] the functor given on objects by left multiplication by $\tau(e)$ is isomorphic to ${}^*R_L^G$. So now it suffices to show that for $V \in \mathrm{mod}(FG)$, the FL-modules $\tau(e)V$ and $(eV^\tau)^\tau$ are naturally isomorphic. Identifying V and $V^{\tau\tau}$, we have the natural isomorphism $(eV^\tau)^\tau \cong V/(eV^\tau)^\circ$ where

$$\begin{aligned}(eV^\tau)^\circ &= \{v \in V \mid (ef)(v) = 0 \text{ for all } f \in V^\tau\}\\ &= \{v \in V \mid f(\tau(e)v) = 0 \text{ for all } f \in V^\tau\} = (1 - \tau(e))V.\end{aligned}$$

We see that $(eV^\tau)^\tau \cong V/((1-\tau(e))V) \cong \tau(e)V$ as required.

Now to show that the functors $(?)^\tau \circ R_L^G \circ (?)^\tau$ and R_L^G are isomorphic, we note that by what we have proved so far and (2.2a), R_L^G is left adjoint to $(?)^\tau \circ {}^*R_L^G \circ (?)^\tau$. So by uniqueness of adjoint functors, it suffices to check that $(?)^\tau \circ R_L^G \circ (?)^\tau$ is also left adjoint to $(?)^\tau \circ {}^*R_L^G \circ (?)^\tau$. For any $U \in \mathrm{mod}(FL), V \in \mathrm{mod}(FG)$, we have

$$\begin{aligned}\mathrm{Hom}_{FG}((R_L^G(U^\tau))^\tau, V) &\cong \mathrm{Hom}_{FG}(V^\tau, R_L^G(U^\tau))\\ &\cong \mathrm{Hom}_{FL}({}^*R_L^G(V^\tau), U^\tau) \cong \mathrm{Hom}_{FL}(U, ({}^*R_L^G(V^\tau))^\tau)\end{aligned}$$

as required for adjointness. □

2.2f. **Corollary.** *Given any irreducible left FL-module M, the FG-module $R_L^G M$ is isomorphic to $(R_L^G M)^\tau$.*

Proof. We have observed that $M \cong M^\tau$. So, $R_L^G M \cong R_L^G(M^\tau) \cong (R_L^G M)^\tau$ by the lemma. □

2.3. Characters and Deligne-Lusztig operators

Let $G = G_n$ and write $\mathbb{Q}_p$ for the p-adic field with algebraic closure $\bar{\mathbb{Q}}_p$. We write $X(G)$ for the character ring of G over $\bar{\mathbb{Q}}_p$, and similarly for any Levi subgroup of G. The irreducible characters in $X(G)$ were originally constructed by Green [G$_1$]. We will adopt for brevity the point of view of Fong and Srinivasan [FS], where Green's results are reformulated in terms of the Deligne-Lusztig theory; see also [LS], [DM$_1$] and [DM$_2$, §15.4] for this point of view. The very elegant approach of [Z] (which contains many but not all of the results we need) might also serve as a useful introduction.

The parametrization of irreducibles described in [FS, §1] depends on first fixing an embedding $\bar{\mathbb{F}}_q^\times \hookrightarrow \bar{\mathbb{Q}}_p^\times$, as well as on the choice of parametrization of the irreducible characters of the symmetric groups by partitions reviewed in §1.1. Having made these choices, we obtain an irreducible character $\chi_{s,\underline{\lambda}} \in X(G)$ for each block-diagonal element $s \in G$ and each $\underline{\lambda} \vdash \kappa(s)$. The irreducible characters in $X(G)$ are the characters $\{\chi_{s,\underline{\lambda}} \mid s \in \mathcal{C}_{ss}, \underline{\lambda} \vdash \kappa(s)\}$.

We say a little more about the construction of $\chi_{s,\underline{\lambda}}$. For a Levi subgroup L of G, $R_L^G : X(L) \to X(G)$ now denotes the Deligne-Lusztig operator as introduced in [DL, L$_1$]. Note we are abusing notation somewhat here: strictly speaking we should denote the operator R_L^G by $R_{\underline{L}}^{\underline{G}}$, where $\underline{G}$ is the algebraic group $GL_n(\bar{\mathbb{F}}_q)$ and $\underline{L}$ is a Levi subgroup of $\underline{G}$ stable under the Frobenius map f_q of (2.1.6), with the Levi subgroup L of G equal to the set of f_q-fixed points in $\underline{L}$.

If L happens to be a standard Levi subgroup of G, we also have operators $R_L^G : X(L) \to X(G)$ and ${}^*R_L^G : X(G) \to X(L)$ induced by the Harish-Chandra induction and restriction functors of §2.2 over $\bar{\mathbb{Q}}_p$. It is known that the Harish-Chandra operator R_L^G coincides with the Deligne-Lusztig operator in this special case that L is standard.

Recall that for $\mu \vdash n$, T_μ denotes a representative of the corresponding conjugacy class of maximal tori of G_n, as in (2.1.7). Also, as in Lemma 1.3f, we write c_μ for the number of elements of Σ_n of cycle-type μ. For $\lambda \vdash n$, the class function

$$\chi_\lambda = \frac{1}{n!} \sum_{\mu \vdash n} c_\mu \phi_\lambda(\mu) R_{T_\mu}^G(1) \tag{2.3.1}$$

is an irreducible *unipotent* character of G. The formula (2.3.1) (which is [FS, (1.13)]) can be inverted to give the formula [FS, (1.14)]:

$$R_{T_\mu}^G(1) = \sum_{\lambda \vdash n} \phi_\lambda(\mu) \chi_\lambda. \tag{2.3.2}$$

For $\lambda = (l_1, \dots, l_a)$ with transpose $\lambda' = (l'_1, \dots, l'_b)$, set

$$m(\lambda) = \sum_{j=1}^{b} (l'_j)^2 - \sum_{i=1}^{a} i l_i.$$

The degree of the unipotent character χ_λ is given by [FS, (1.15)]:

$$\chi_\lambda(1) = \frac{q^{m(\lambda)}(q^n - 1)(q^{n-1} - 1)\dots(q - 1)}{\prod_h (q^h - 1)} \tag{2.3.3}$$

where h runs over the hook lengths of λ.

Suppose now in general that $s \in G$ is a block-diagonal element of the form (2.1.1). The choice of embedding $\bar{\mathbb{F}}_q^\times \hookrightarrow \bar{\mathbb{Q}}_p^\times$ allows us to associate to s a linear character $\hat{s} \in X(C_G(s))$, as in [FS, (1.16)]. Recalling the structure of $C_G(s)$ from (2.1.5), we have for each $\underline{\lambda} = (\lambda_1, \dots, \lambda_a) \vdash \kappa(s)$ the irreducible unipotent character $\chi_{\lambda_1} \cdots \chi_{\lambda_a}$ (outer product) of $C_G(s)$. Note we are abusing notation here (and later) by allowing q to vary to give the definition of the unipotent character $\chi_{\lambda_i} \in X(GL_{k_i}(q^{d_i}))$, $i = 1, \dots, a$. Define the sign ε_s by

$$\varepsilon_s = (-1)^{n + k_1 + \dots + k_a}. \tag{2.3.4}$$

Then,

$$\chi_{s,\underline{\lambda}} = \varepsilon_s R^G_{C_G(s)}(\hat{s}.\chi_{\lambda_1} \cdots \chi_{\lambda_a}) \tag{2.3.5}$$

is precisely the irreducible character of G parametrized by the pair $(s, \underline{\lambda})$.

Using [FS, (1.3)] (cf. [DM$_1$, 3.10]), we see that the degree of $\chi_{s,\underline{\lambda}}$ can be computed from (2.3.3) by the formula:

$$\chi_{s,\underline{\lambda}}(1) = |G : C_G(s)|_{q'} \chi_{\lambda_1}(1) \dots \chi_{\lambda_a}(1), \tag{2.3.6}$$

where for an integer N, $N_{q'}$ denotes its largest divisor coprime to q. We repeat: in calculating the degree $\chi_{\lambda_i}(1)$ using (2.3.3) q needs to be replaced by q^{d_i}, since $\chi_{\lambda_i} \in X(GL_{k_i}(q^{d_i}))$ not $X(GL_n(q))$ as there.

The following property, proved in [FS, p. 140], is of great importance to the modular theory:

(2.3a) *Let $t = sy$ be a block-diagonal element of G with p-regular part s. For any $\underline{\lambda} = (\lambda_1, \dots, \lambda_b) \vdash \kappa(t)$, the generalized characters $R^G_{C_G(t)}(\hat{t}.\chi_{\lambda_1} \cdots \chi_{\lambda_b})$ and $R^G_{C_G(s)}(\hat{s}.R^{C_G(s)}_{C_G(t)}(\chi_{\lambda_1} \cdots \chi_{\lambda_b}))$ agree on all p-regular conjugacy classes of G.*

As our first application of (2.3a), we have:

2.3b. **Lemma.** *With $t = sy$ and $\underline{\lambda}$ as in (2.3a), the Brauer character obtained by restricting the character $\chi_{t,\underline{\lambda}}$ to the p-regular classes of G can be written as a $\mathbb{Z}$-linear combination of the restrictions of the characters $\{\chi_{s,\underline{\mu}} \mid \underline{\mu} \vdash \kappa(s)\}$ to p-regular classes.*

Proof. Using (2.3a), we just need to observe that all of the irreducible constituents of $R_{C_G(t)}^{C_G(s)}(\chi_{\lambda_1}\cdots\chi_{\lambda_b})$ are unipotent characters of $C_G(s)$. Let T be a maximal split torus of $C_G(t)$. Then, $\chi_{\lambda_1}\cdots\chi_{\lambda_b}$ is a constituent of $R_T^{C_G(t)}(1)$. So it suffices by transitivity of Deligne-Lusztig operators to observe that all constituents of $R_T^{C_G(s)}(1)$ are unipotent characters of $C_G(s)$, which follows by (2.3.2) (noting that T is also a maximal torus of $C_G(s)$). □

Now fix $\sigma \in \bar{\mathbb{F}}_q^\times$ of degree d over $\mathbb{F}_q$ and suppose that $n = kd$ for some $k \geq 1$. Abbreviate $\chi_{s,\underline{\lambda}}$ in the special case $s = (\sigma)^k$ by $\chi_{\sigma,\lambda}$, where $\underline{\lambda} = (\lambda)$ for $\lambda \vdash k$. Recall again that c_λ denotes the number of elements of Σ_k of cycle-type λ. The second fundamental consequence of (2.3a) (cf. Lemma 1.3f and [DJ$_2$, Lemma 3.2]) is as follows:

2.3c. **Lemma.** *Let $\tau \in \bar{\mathbb{F}}_q^\times$ be of degree $e = md$ over $\mathbb{F}_q$, with p-regular part conjugate to σ. Then, for any $l \geq 1$, $k = ml$ and $n = kd = le$, the character $\chi_{\tau,(l)}$ agrees with the generalized character*

$$\frac{(-1)^{k+l}}{l!}\sum_{\lambda\vdash l}\sum_{\mu\vdash k} c_\lambda \phi_\mu(m\lambda)\chi_{\sigma,\mu}$$

on all p-regular classes of G_n.

Proof. Let $t = (\tau)^l$ and $s = (\sigma)^k$, so that $C_G(t) \cong GL_l(q^e)$ and $C_G(s) \cong GL_k(q^d)$. By definition, $R_{C_G(t)}^G(\hat{t}.\chi_{(l)}) = (-1)^{n+l}\chi_{\tau,(l)}$ and $R_{C_G(s)}^G(\hat{s}.\chi_\mu) = (-1)^{n+k}\chi_{\sigma,\mu}$ for any $\mu \vdash k$. So applying (2.3a), it suffices to show that

$$R_{C_G(t)}^{C_G(s)}\chi_{(l)} = \frac{1}{l!}\sum_{\lambda\vdash l}\sum_{\mu\vdash k} c_\lambda\phi_\mu(m\lambda)\chi_\mu.$$

Now, $\phi_{(l)}$ is the trivial character of the symmetric group, so by (2.3.1), we can write the unipotent character $\chi_{(l)}$ of $C_G(t) \cong GL_l(q^e)$ as

$$\chi_{(l)} = \frac{1}{l!}\sum_{\lambda\vdash l} c_\lambda R_{T_\lambda}^{C_G(t)}(1)$$

where $T_\lambda \leq C_G(t)$ is a maximal torus isomorphic to $GL_1(q^{el_1}) \times \cdots \times GL_1(q^{el_a})$ for $\lambda = (l_1, \ldots, l_a) \vdash l$. Observe that T_λ is also a maximal torus of $C_G(s)$; we should then denote it instead by $T_{m\lambda} \leq C_G(s)$. Now, using transitivity of the Deligne-Lusztig operators, we deduce from (2.3.2) that

$$R_{C_G(t)}^{C_G(s)}\chi_{(l)} = \frac{1}{l!}\sum_{\lambda\vdash l} c_\lambda R_{T_{m\lambda}}^{C_G(s)}(1) = \frac{1}{l!}\sum_{\lambda\vdash l}\sum_{\mu\vdash k} c_\lambda\phi_\mu(m\lambda)\chi_\mu$$

as required to complete the proof. □

By transitivity of Deligne-Lusztig induction, we see immediately from the definition (2.3.5) that:

(2.3d) *Suppose that s and t are block-diagonal element of G_n with $C_{G_n}(t) \subseteq C_{G_n}(s)$. For $\underline{\lambda} = (\lambda_1, \dots, \lambda_a) \vdash \kappa(t)$, $\chi_{t,\underline{\lambda}} = R^G_{G_{\pi(s)}} \chi^{\mathrm{Levi}}_{t,\underline{\lambda}}$ where $\chi^{\mathrm{Levi}}_{t,\underline{\lambda}}$ denotes the irreducible character $\varepsilon_t R^{G_{\pi(s)}}_{C_G(t)}(\hat{t}.\chi_{\lambda_1} \cdots \chi_{\lambda_a})$ of $G_{\pi(s)}$.*

As a special case of this (taking $s = t$), we see that an arbitrary irreducible character $\chi_{s,\underline{\lambda}}$ can easily be expressed in terms of the $\chi_{\sigma,\lambda}$'s:

(2.3e) *For a block-diagonal $s \in G_n$ of the form (2.1.1) and $\underline{\lambda} = (\lambda_1, \dots, \lambda_a) \vdash \kappa(s)$, $\chi_{s,\underline{\lambda}} = R^{G_n}_{G_{\pi(s)}}(\chi_{\sigma_1,\lambda_1} \cdots \chi_{\sigma_a,\lambda_a})$.*

In view of (2.3e), it will simplify notation in what follows to always restrict our attention initially to characters of the form $\chi_{\sigma,\lambda}$. Now that we have this notation in place, we will not need to use the Deligne-Lusztig operators again.

We record the following formula for the degree of $\chi_{\sigma,\lambda}$, which follows easily from (2.3.3) and (2.3.6):

$$\chi_{\sigma,\lambda}(1) = \frac{q^{dm(\lambda)}(q^n - 1)(q^{n-1} - 1) \dots (q - 1)}{\prod_h (q^{dh} - 1)} \tag{2.3.7}$$

where h runs over the hook lengths of λ. We will also use the following inner product formula which is a special case of [FS, (1B), (1C)]:

(2.3f) *Fix $\nu = (k_1, \dots, k_a) \vDash k$ and a multi-partition $\underline{\lambda} = (\lambda_1, \dots, \lambda_a) \vdash \nu$. Then, every irreducible constituent of $R^{G_n}_{G_{d\nu}} \chi_{\sigma,\lambda_1} \cdots \chi_{\sigma,\lambda_a}$ is of the form $\chi_{\sigma,\mu}$ for $\mu \vdash k$. Moreover, given $\mu \vdash k$,*

$$(\chi_{\sigma,\mu}, R^{G_n}_{G_{d\nu}} \chi_{\sigma,\lambda_1} \cdots \chi_{\sigma,\lambda_a}) = (\phi_\mu, \mathrm{ind}^{\Sigma_k}_{\Sigma_\nu} \phi_{\lambda_1} \cdots \phi_{\lambda_a})$$

where the inner products are the usual ones on $X(G_n)$ and $X(\Sigma_k)$ respectively.

2.4. Cuspidal representations and blocks

If G is a standard Levi subgroup of G_n, a *cuspidal FG-module* is an FG-module M with the property that ${}^*R^G_L M = 0$ for all standard Levi subgroups $L < G$. Similarly, a *cuspidal character* $\chi \in X(G)$ is a character with the property that ${}^*R^G_L \chi = 0$ for all standard Levi subgroups $L < G$.

The irreducible cuspidal characters of G_n are precisely the $\chi_{\sigma,(1)}$ for $\sigma \in \bar{\mathbb{F}}_q^\times$ of degree n over $\mathbb{F}_q$. We will write simply χ_σ for the cuspidal character $\chi_{\sigma,(1)}$ for such σ. We fix a KG_n-module $M(\sigma)_K$ affording the character χ_σ, for each σ. Choose an $\mathcal{O}$-free $\mathcal{O}G_n$-module $M(\sigma)_\mathcal{O}$ such that $M(\sigma)_K \cong K \otimes_\mathcal{O} M(\sigma)_\mathcal{O}$, and set $M(\sigma) = M(\sigma)_F = F \otimes_\mathcal{O} M(\sigma)_\mathcal{O}$. Since Harish-Chandra restriction commutes with base change, $M(\sigma)$ is a cuspidal FG_n-module. We will later see (cf. Lemma 2.5f and Theorem 4.3b) that $M(\sigma)$ is actually an irreducible FG_n-module, so that its definition does not depend on the particular choice of the $\mathcal{O}$-lattice $M(\sigma)_{\mathcal{O}}$, but this is not yet clear. We just note for now that by (2.3.7),

$$\dim M(\sigma) = (q-1)(q^2-1)\dots(q^{d-1}-1), \tag{2.4.1}$$

interpreted as 1 if $d=1$.

Now let R be one of the rings F, K or $\mathcal{O}$. For $\sigma \in \bar{\mathbb{F}}_q^\times$ of degree d over $\mathbb{F}_q$ and $n = kd$, the module $M(\sigma)_R \boxtimes \cdots \boxtimes M(\sigma)_R$, that is, the outer tensor product of k copies of $M(\sigma)_R$, is a cuspidal $RG_{(d^k)}$-module. We define

$$M^k(\sigma)_R = R^{G_n}_{G_{(d^k)}} M(\sigma)_R \boxtimes \cdots \boxtimes M(\sigma)_R. \tag{2.4.2}$$

Since Harish-Chandra induction commutes with base change, $M^k(\sigma)_{\mathcal{O}}$ is an $\mathcal{O}$-lattice in $M^k(\sigma)_K$, and $M^k(\sigma) := M^k(\sigma)_F \cong F \otimes_{\mathcal{O}} M^k(\sigma)_{\mathcal{O}}$.

Observe that $M^k(\sigma)_K$ is a module affording the character $R^{G_n}_{G_{(d^k)}} \chi_\sigma \cdots \chi_\sigma$ (k times). So according to (2.3f), the irreducible constituents of $M^k(\sigma)_K$ have characters among $\{\chi_{\sigma,\lambda} \,|\, \lambda \vdash k\}$. Moreover, by (2.3f) again, $(\chi_{\sigma,\lambda}, R^{G_n}_{G_{(d^k)}} \chi_\sigma \cdots \chi_\sigma) = \phi_\lambda(1)$. Since the only one dimensional characters of the symmetric group are the trivial and sign characters $\phi_{(k)}$ and $\phi_{(1^k)}$ respectively, we deduce:

(2.4a) *The irreducible constituents of $M^k(\sigma)_K$ have characters $\{\chi_{\sigma,\lambda} \mid \lambda \vdash k\}$, and the only constituents appearing with multiplicity one are those with characters $\chi_{\sigma,(k)}$ and $\chi_{\sigma,(1^k)}$.*

More generally, for a block-diagonal element s as in (2.1.1), set

$$M(s)_R = R^{G_n}_{G_{\pi(s)}} M^{k_1}(\sigma_1)_R \boxtimes \cdots \boxtimes M^{k_a}(\sigma_a)_R \tag{2.4.3}$$

and write simply $M(s)$ for $M(s)_F$. By transitivity of Harish-Chandra induction, $M(s)_R$ is precisely the RG_n-module Harish-Chandra induced from the cuspidal $RG_{\delta(s)}$-module

$$\overbrace{M(\sigma_1)_R \boxtimes \cdots \boxtimes M(\sigma_1)_R}^{k_1 \text{ times}} \boxtimes \cdots \boxtimes \overbrace{M(\sigma_a)_R \boxtimes \cdots \boxtimes M(\sigma_a)_R}^{k_a \text{ times}}. \tag{2.4.4}$$

Applying (2.4a) and (2.3e), we see immediately that the irreducible constituents of $M(s)_K$ in characteristic 0 correspond to the irreducible characters $\{\chi_{s,\underline{\lambda}} \mid \underline{\lambda} \vdash \kappa(s)\}$. These irreducible constituents for fixed s constitute the *geometric conjugacy class* of irreducible KG_n-modules corresponding to s.

Now we can state a fundamental result from block theory, which follows as a consequence of the classification of blocks obtained by Fong and Srinavasan [FS]. It also has a short direct proof, valid for arbitrary type and independent of the full block classification of *loc. cit.*, due to Broué and Michel [BM].

(2.4b) *Suppose s and t are block-diagonal elements of G_n such that the p-regular parts of s and t are not conjugate in G_n. Then, the irreducible characters $\chi_{s,\underline{\lambda}}$ and $\chi_{t,\underline{\mu}}$ belong to different p-blocks, for all $\underline{\lambda} \vdash \kappa(s), \underline{\mu} \vdash \kappa(t)$.*

Let s be a p-regular block-diagonal element of G_n. By (2.4b), we can find a central idempotent $e_s \in \mathcal{O}G_n$ such that e_s acts as the identity on $M(t)_K$ for all block-diagonal $t \in G_n$ with p-regular part conjugate to s and as zero on $M(t)_K$ for all other block-diagonal $t \in G_n$. The set $\{e_s \mid s \in \mathcal{C}_{ss,p'}\}$ is a set of mutually orthogonal central idempotents in $\mathcal{O}G_n$ summing to the identity. For $R = K, F$ or $\mathcal{O}$, let $B_{s,R}$ be the union of blocks of RG_n corresponding to the central idempotent e_s, and write simply B_s for the algebra $B_{s,F}$. So, the set

$$\bigcup_t \{\chi_{t,\underline{\lambda}} \mid \underline{\lambda} \vdash \kappa(t)\}, \tag{2.4.5}$$

as t runs over all elements of $\mathcal{C}_{ss}$ with p-regular part conjugate to s, gives a complete set of non-isomorphic irreducible characters belonging to the block $B_{s,K}$. Then,

$$RG_n = \bigoplus_{s \in \mathcal{C}_{ss,p'}} B_{s,R} \tag{2.4.6}$$

is a decomposition of the group algebra as a direct sum of two-sided ideals. Let Par_k denote the number of partitions of k, and more generally for a composition $\kappa = (k_1, \dots, k_a)$ write Par_κ for the number of multi-partitions of κ, so $\mathrm{Par}_\kappa = \mathrm{Par}_{k_1} \dots \mathrm{Par}_{k_a}$. We have (see also [GH, §3]):

2.4c. **Lemma.** *For a p-regular block-diagonal $s \in G_n$, the F-algebra B_s has precisely* $\mathrm{Par}_{\kappa(s)}$ *non-isomorphic irreducible modules, all of which appear as constituents of $M(s)$. In particular, for a p-regular $\sigma \in \bar{\mathbb{F}}_q^\times$, $M^k(\sigma)$ has precisely* Par_k *non-isomorphic composition factors.*

Proof. We may assume that $s \in \mathcal{C}_{ss,p'}$. Using (2.4.5), the ring of Brauer characters of B_s is spanned by the restrictions to p-regular classes of the characters $\bigcup_t \{\chi_{t,\underline{\lambda}} \mid \underline{\lambda} \vdash \kappa(t)\}$ as t runs over all elements of $\mathcal{C}_{ss}$ with p-regular part conjugate to s.

Given this, Lemma 2.3b implies in fact that the ring of Brauer characters of B_s is spanned just by the restrictions of the $\{\chi_{s,\underline{\mu}} \mid \underline{\mu} \vdash \kappa(s)\}$. Hence, the algebra B_s has at most $\mathrm{Par}_{\kappa(s)}$ distinct irreducibles, for all $s \in \mathcal{C}_{ss,p'}$. The total number of irreducible FG_n-modules is equal to the number of p-regular conjugacy classes, so now (2.4.6) and an elementary counting argument shows that in fact B_s has precisely $\mathrm{Par}_{\kappa(s)}$ distinct irreducibles.

We have now shown that the Brauer characters obtained by restricting $\{\chi_{s,\underline{\lambda}} \mid \underline{\lambda} \vdash \kappa(s)\}$ to p-regular classes give a basis for the ring of Brauer characters of B_s. Since each $\chi_{s,\underline{\lambda}}$ appears as a constituent of $M(s)_K$, it follows that all of the $\mathrm{Par}_{\kappa(s)}$ distinct irreducible B_s-modules definitely appear as composition factors of $M(s)$. □

We will also need to consider blocks of Levi subgroups of G_n in one special case. So fix now a p-regular block-diagonal element $s \in G_n$, suppose that $\pi(s) = (n_1, \dots, n_a) \models n$ is defined as in (2.1.4), and write $s = s_1 \dots s_a$ with s_i lying in the factor G_{n_i} of $G_{\pi(s)}$. For $R = F, K$ or $\mathcal{O}$, define $B_{s,R}^{\text{Levi}}$ to be the block $B_{s_1,R} \otimes \dots \otimes B_{s_a,R}$ of $RG_{\pi(s)} \cong RG_{n_1} \otimes \dots \otimes RG_{n_a}$.

2.4d. **Lemma.** *With notation as above, the Harish-Chandra operator $R_{G_{\pi(s)}}^{G_n}$ gives a bijection between the set of irreducible characters of $B_{s,K}^{\text{Levi}}$ and the set of irreducible characters of $B_{s,K}$.*

Proof. If t is a block-diagonal element of G_n with p-regular part s, we observe that $C_{G_n}(t) \subseteq C_{G_n}(s) \subseteq G_{\pi(s)}$. For $\underline{\lambda} = (\lambda_1, \dots, \lambda_a) \vdash \kappa(t)$, let $\chi_{t,\underline{\lambda}}^{\text{Levi}}$ denote the irreducible character $\varepsilon_t R_{C_G(t)}^{G_{\pi(s)}}(\hat{t}.\chi_{\lambda_1} \cdots \chi_{\lambda_a})$ of $G_{\pi(s)}$ as in (2.3d). Then, the irreducible characters of $B_{s,K}$ (resp. $B_{s,K}^{\text{Levi}}$) are precisely the characters $\chi_{t,\underline{\lambda}}$ (resp. $\chi_{t,\underline{\lambda}}^{\text{Levi}}$) for all $\underline{\lambda} \vdash \kappa(t)$, as t runs over a set of representatives of the block-diagonal elements $t \in G_n$ with p-regular part equal to s. So the lemma follows directly from (2.3d). □

2.4e. **Theorem.** *With notation as above and $R = F, K$ or $\mathcal{O}$, the Harish-Chandra induction functor $R_{G_{\pi(s)}}^{G_n}$ induces a Morita equivalence between $B_{s,R}^{\text{Levi}}$ and $B_{s,R}$*

Proof. We just need to prove this in the case $R = \mathcal{O}$, the other cases following immediately from this since Harish-Chandra induction commutes with base change. Let us first observe that $R_{G_{\pi(s)}}^{G_n}$ does indeed restrict to a well-defined functor from $B_{s,\mathcal{O}}^{\text{Levi}}$ to $B_{s,\mathcal{O}}$. Thanks to Lemma 2.4d, it certainly sends torsion free $B_{s,\mathcal{O}}^{\text{Levi}}$-modules to $B_{s,\mathcal{O}}$-modules. But an arbitrary $B_{s,\mathcal{O}}^{\text{Levi}}$-module M is a quotient of a projective, which is torsion free, so by exactness, $R_{G_{\pi(s)}}^{G_n} M$ is a quotient of a $B_{s,\mathcal{O}}$-module, hence itself a $B_{s,\mathcal{O}}$-module.

Now let $G = G_n$ and $L = G_{\pi(s)}$. Denote the standard parabolic subgroup of G with Levi factor L by P. Let X be a projective generator for $\text{mod}(B_{s,\mathcal{O}}^{\text{Levi}})$ and set $Y = R_L^G(X) = \mathcal{O}G \otimes_{\mathcal{O}P} X$. We will also write $KX = K \otimes_{\mathcal{O}} X$, $KY = K \otimes_{\mathcal{O}} Y$ and identify KY with $R_L^G(KX) = KG \otimes_{KP} KX$. In view of the previous paragraph and (2.2a), Y is a projective $B_{s,\mathcal{O}}$-module.

Writing endomorphisms on the right, consider the endomorphism algebras

$$E^{\text{Levi}} = \text{End}_{B_{s,\mathcal{O}}^{\text{Levi}}}(X), \qquad E = \text{End}_{B_{s,\mathcal{O}}}(Y).$$

The functor R_L^G determines an algebra homomorphism $\theta : E^{\text{Levi}} \to E$, which is clearly injective. We claim that θ is surjective, hence an isomorphism. Applying Lemma 2.4d, every irreducible constituent of KX is mapped to an irreducible constituent of KY and multiplicities are preserved. Consequently, comparing dimensions, R_L^G certainly induces an algebra isomorphism from $\text{End}_{KL}(KX)$

to $\mathrm{End}_{KG}(KY)$. In particular, every KG-endomorphism of KY preserves the (KP-direct) summand $1 \otimes_{KP} KX$ of $KY = KG \otimes_{KP} KX$. We conclude that every $\mathcal{O}G$-endomorphism of $Y = R_L^G(X)$ also preserves the direct summand $1 \otimes_{\mathcal{O}P} X$ of Y, since it preserves Y and the space $1 \otimes_{KP} KX$ when extended to an endomorphism of KY. In other words, every $\mathcal{O}G$-endomorphism of Y is induced by an $\mathcal{O}L$-endomorphism of X, that is, θ is surjective as claimed.

As a consequence, there is a one-to-one correspondence between the primitive idempotents in the endomorphism algebras E^{Levi} and E. We deduce that R_L^G takes projective indecomposable summands of X to projective indecomposable summands of Y, and preserves distinct isomorphism types. Applying Lemma 2.4c, the algebras $B_{s,F}^{\mathrm{Levi}}$ and $B_{s,F}$ have the same number of non-isomorphic irreducibles, so $B_{s,\mathcal{O}}^{\mathrm{Levi}}$ and $B_{s,\mathcal{O}}$ have the same number of non-isomorphic projective indecomposables. So since X was a projective generator, Y must also be a projective generator for $\mathrm{mod}(B_{s,\mathcal{O}})$. So now the fact that the endomorphism algebras E^{Levi} and E are isomorphic proved in the previous paragraph means that the algebras $B_{s,\mathcal{O}}^{\mathrm{Levi}}$ and $B_{s,\mathcal{O}}$ are Morita equivalent.

It remains to check that the functor R_L^G itself gives the Morita equivalence. For this, we will identifying E^{Levi} and E via the isomorphism θ. Then, we can regard X (resp. Y) as a $(B_{s,\mathcal{O}}^{\mathrm{Levi}}, E)$-bimodule (resp. a $(B_{s,\mathcal{O}}, E)$-bimodule). We have a diagram of functors:

$$\begin{array}{ccc} \mathrm{mod}(B_{s,\mathcal{O}}^{\mathrm{Levi}}) & \xrightarrow{R_L^G} & \mathrm{mod}(B_{s,\mathcal{O}}) \\ {\scriptstyle X\otimes_E ?}\uparrow & & \uparrow{\scriptstyle Y\otimes_E ?} \\ \mathrm{mod}(E) & = & \mathrm{mod}(E) \end{array}$$

where the vertical functors are Morita equivalences since X and Y are projective generators. Moreover, the diagram commutes, that is, the functors $R_L^G \circ (X \otimes_E ?)$ and $(R_L^G X) \otimes_E ?$ are isomorphic, which follows directly by associativity of tensor product. So the top functor, R_L^G, is also a Morita equivalence. □

Theorem 2.4e plays the role of (2.3e) in the modular theory: it allows us almost all of the time to restrict our attention to studying the blocks of $B_{s,R}$ with s of the form $(\sigma)^k$ for p-regular $\sigma \in \bar{\mathbb{F}}_q^\times$, rather than the more general $B_{s,R}$ for *arbitrary* $s \in \mathcal{C}_{ss,p'}$. The following notation will be convenient: given p-regular $\sigma \in \bar{\mathbb{F}}_q^\times$ and $k \geq 1$, let $B_{\sigma,k,R}$ denote the algebra $B_{s,R}$ in the special case that $s = (\sigma)^k$. Write simply $B_{\sigma,k}$ for $B_{\sigma,k,F}$. We remark at this point that as a special case of a conjecture of Broué [B, p. 61], it is expected that the algebra $B_{\sigma,k,\mathcal{O}}$ is Morita equivalent to the 'unipotent block' $B_{1,\mathcal{O}}$ of $GL_k(\mathbb{F}_{q^d})$, where d is the degree of σ over $\mathbb{F}_q$.

2.5. Howlett-Lehrer theory and the Gelfand-Graev representation

Suppose now that $\sigma \in \bar{\mathbb{F}}_q^\times$ is of degree d over $\mathbb{F}_q$ and that $n = kd$ for some $k \geq 1$. Let R denote one of the rings F, K or $\mathcal{O}$ throughout the section. Recalling the definition of the Hecke algebra from §1.1, we let $H_{k,R} = H_{R,q^d}(\Sigma_k)$. Applying (1.1.1), we may identify $H_{k,F}$ (resp. $H_{k,K}$) with the algebra $F \otimes_{\mathcal{O}} H_{k,\mathcal{O}}$ (resp.

$K \otimes_{\mathcal{O}} H_{k,\mathcal{O}})$. We will usually write simply H_k in place of $H_{k,F}$ over the modular field. We wish to construct a right action of $H_{k,R}$ on the module $M^k(\sigma)_R$, as a very special case of the theory of Howlett and Lehrer [HL$_1$].

Write $N_R = M(\sigma)_R \boxtimes \cdots \boxtimes M(\sigma)_R$ (k times) and $M_R = M^k(\sigma)_R = R^{G_n}_{G_{(d^k)}} N_R$ for short. We identify M_R with $RG_n \otimes_{RP} N_R$, where P is the standard parabolic subgroup of G_n with Levi factor $G_{(d^k)}$ and unipotent radical $Y_{(d^k)}$. Let f denote the idempotent

$$f = \frac{1}{|Y_{(d^k)}|} \sum_{u \in Y_{(d^k)}} u \in RP$$

so that ${}^*R^{G_n}_{G_{(d^k)}} M_R = f M_R$. Finally, recall the embedding $\pi : \Sigma_k \hookrightarrow G_n$ as $d \times d$-block permutation matrices, from §2.1.

2.5a. **Lemma.** $f M_R = \bigoplus_{x \in \Sigma_k} f\pi_x \otimes N_R$

Proof. One argues using the Bruhat decomposition as in [D$_1$, Lemma 3.4] to show that the sum

$$\sum_{x \in \Sigma_k} f\pi_x \otimes N_R \subseteq f M_R$$

is direct and has a complement in $f M_R$ as an R-module. It then just remains to check that the R-rank of $f M_R$ is $k!(\operatorname{rank} N_R)$. But by the Mackey decomposition, writing $L = G_{(d^k)}$,

$$f M_R = {}^*R^{G_n}_L M_R \cong \bigoplus_{w \in D_{(d^k),(d^k)}} R^L_{L \cap {}^w L} \circ \operatorname{conj}_w \circ {}^*R^L_{L \cap {}^{w^{-1}} L}(N_R).$$

Since N_R is a cuspidal RL-module, the summand corresponding to w is zero unless $L \cap {}^{w^{-1}} L = L$, in which case according to (2.1c) $w = \pi_x$ for $x \in \Sigma_k$. So, $f M_R \cong \bigoplus_{x \in \Sigma_k} \operatorname{conj}_{\pi_x}(N_R)$ which has the required R-rank. □

Notice that each summand $f\pi_w \otimes N_R$ appearing in the lemma is an $RG_{(d^k)}$-submodule of $f M_R$. Moreover, given $w \in \Sigma_k$, there is an isomorphism of $RG_{(d^k)}$-modules $\bar{A}_w : N_R \to f\pi_w \otimes N_R$, such that

$$v_1 \otimes \cdots \otimes v_k \mapsto f\pi_w \otimes v_{w1} \otimes \cdots \otimes v_{wk}$$

for all $v_1, \dots, v_k \in M(\sigma)_R$. Identifying ${}^*R^{G_n}_{G_{(d^k)}} M_R$ with $\bigoplus_{x \in \Sigma_k} f\pi_x \otimes N_R$ according to Lemma 2.5a, we extend the range of $\bar{A}_w$ to obtain an $RG_{(d^k)}$-module monomorphism

$$A_w : N_R \to {}^*R^{G_n}_{G_{(d^k)}} M_R, \tag{2.5.1}$$

for each $w \in \Sigma_k$. Clearly, the maps $\{A_w \mid w \in \Sigma_k\}$ are linearly independent. Now for each $w \in \Sigma_k$, let

$$B_w : M_R \to M_R$$

be the RG_n-module homomorphism induced by A_w under the isomorphism

$$\mathrm{Hom}_{RG_{(d^k)}}(N_R, {}^*R^{G_n}_{G_{(d^k)}} M_R) \cong \mathrm{Hom}_{RG_n}(M_R, M_R)$$

arising from adjointness (2.2a). So, we have constructed $k!$ linearly independent RG_n-endomorphisms of M_R, namely, $\{B_w \mid w \in \Sigma_k\}$. The Howlett-Lehrer theory [HL$_1$] in our special case (see [D$_1$, Lemma 3.5] and [J$_2$, Theorem 4.12]) computes relations between the B_w (observe that knowing a relation over K implies the relation over $\mathcal{O}$, hence over F) to show:

(2.5b) *There is an algebra embedding $\theta : H_{k,R} \hookrightarrow \mathrm{End}_{RG_n}(M_R)$ (writing endomorphisms on the right) with $\theta(T_w) = ((-1)^{o(\sigma)+1} q^{\frac{1}{2}d(d+1)})^{\ell(w)} B_w$ for all $w \in \Sigma_k$, where $o(\sigma)$ is the order of $\sigma \in \bar{\mathbb{F}}_q^\times$. In the case $R = K$, θ is an isomorphism.*

Later on we will show that θ is an isomorphism for $R = F, \mathcal{O}$ too, but we cannot prove this yet. In view of (2.5b), we will henceforth always regard $M^k(\sigma)_R$ as an $(RG_n, H_{k,R})$-bimodule; it is clear from (2.5b) that the bimodule structure is compatible with base change.

Next, we focus on the case $R = K$. Then, $M^k(\sigma)_K$ is a completely reducible KG_n-module and its endomorphism algebra

$$H_{k,K} \cong \mathrm{End}_{KG_n}(M^k(\sigma)_K)$$

is semisimple. Fitting's lemma gives a bijection between the irreducible KG_n-modules appearing as constituents of $M^k(\sigma)_K$ and the irreducible $H_{k,K}$-modules. We are interested here in the constituents of $M^k(\sigma)_K$ corresponding to the trivial representation $\mathcal{I}_{H_{k,K}}$ and the sign representation $\mathcal{E}_{H_{k,K}}$ of $H_{k,K}$, as defined in §1.1. These appear in $M^k(\sigma)_K$ with multiplicity one since $\mathcal{I}_{H_{k,K}}$ and $\mathcal{E}_{H_{k,K}}$ are one dimensional. Over K, (1.1b) easily implies that x_k and y_k are idempotents up to multiplication by non-zero scalars, so these constituents of $M^k(\sigma)_K$ are according to Fitting's lemma precisely the irreducible submodules $M^k(\sigma)_K x_k$ and $M^k(\sigma)_K y_k$, respectively.

2.5c. **Lemma.** *The KG_n-module $M^k(\sigma)_K x_k$ (resp. $M^k(\sigma)_K y_k$) corresponds to the irreducible character $\chi_{\sigma,(k)}$ (resp. $\chi_{\sigma,(1^k)}$).*

Proof. We may assume $k > 1$, the case $k = 1$ being trivial. Note that according to (2.3.7), we know the degrees:

$$\chi_{\sigma,(k)}(1) = \prod_{i=1}^{kd}(q^i - 1) \Big/ \prod_{i=1}^{k}(q^{id} - 1), \tag{2.5.2}$$

$$\chi_{\sigma,(1^k)}(1) = q^{\frac{d}{2}k(k-1)} \prod_{i=1}^{kd}(q^i - 1) \Big/ \prod_{i=1}^{k}(q^{id} - 1). \tag{2.5.3}$$

For $k > 1$ these degrees are different. Moreover, by Fitting's lemma, the only irreducible constituents of $M^k(\sigma)_K$ appearing with multiplicity one are the modules $M^k(\sigma)_K x_k$ and $M^k(\sigma)_K y_k$. So applying (2.4a), we see that it suffices to check that $\dim M^k(\sigma)_K x_k = \chi_{\sigma,(k)}(1)$ *not* $\chi_{\sigma,(1^k)}(1)$.

The dimension of $M(\sigma)_K$ is given by (2.4.1). Using this, an easy calculation gives the dimension of the induced module:

$$\dim M^k(\sigma)_K = \prod_{i=1}^{kd}(q^i - 1) \Big/ (q^d - 1)^k.$$

Now [C, Proposition 10.9.6] tells us that

$$\dim M^k(\sigma)_K x_k = \dim M^k(\sigma)_K \Big/ \sum_{w \in \Sigma_k} (q^d)^{\ell(w)}.$$

It is well-known that $\sum_{w \in \Sigma_k} t^{\ell(w)} = \prod_{i=1}^k (t^i - 1)/(t-1)^k$, and one easily verifies now that $\dim M^k(\sigma)_K x_k = \chi_{\sigma,(k)}(1)$. □

We conclude the chapter by reviewing the fundamental properties of the Gelfand-Graev representation introduced in [GG]. Fix a non-trivial homomorphism

$$\chi_K : (\mathbb{F}_q, +) \to K^\times. \tag{2.5.4}$$

Note that the values of χ_K lie in $\mathcal{O}$, so we can restrict χ_K to $\chi_{\mathcal{O}} : (\mathbb{F}_q, +) \to \mathcal{O}^\times$, then reduce module p to obtain the non-trivial character $\chi = \chi_F : (\mathbb{F}_q, +) \to F^\times$.

For $u \in U_n$ and $R = F, K$ or $\mathcal{O}$, let

$$\theta_{n,R}(u) = \chi_R(\sum_{i=1}^{n-1} u_{i,i+1})$$

where $u_{i,i+1}$ denotes the $(i, i+1)$-entry of the matrix u. We associate to this linear character of RU_n the idempotent

$$\gamma_{n,R} = \frac{1}{|U_n|} \sum_{u \in U_n} \theta_{n,R}(u^{-1}) u \in RG_n. \tag{2.5.5}$$

In particular, $\gamma_n := \gamma_{n,F}$. The left ideal $FG_n\gamma_n$ is the *Gelfand-Graev representation* Γ_n of FG_n. Observe that Γ_n is a projective FG_n-module. Moreover, it is the reduction modulo p of $\Gamma_{n,K} = KG_n\gamma_{n,K}$ via the $\mathcal{O}$-lattice $\Gamma_{n,\mathcal{O}} = \mathcal{O}G_n\gamma_{n,\mathcal{O}}$.

More generally, given $\nu = (n_1, \dots, n_a) \vDash n$, we have the analogous left FG_ν-module denoted Γ_ν, which is just the outer tensor product $\Gamma_{n_1} \boxtimes \cdots \boxtimes \Gamma_{n_a}$. This is the left ideal $FG_\nu\gamma_\nu$ where

$$\gamma_\nu = \gamma_{n_1} \otimes \cdots \otimes \gamma_{n_a} \in FG_\nu = FG_{n_1} \otimes \cdots \otimes FG_{n_a}.$$

We also write $\Gamma_{\nu,\mathcal{O}}$ and $\Gamma_{\nu,K}$ for the Gelfand-Graev representations over $\mathcal{O}$ and K, defined in the same way using the idempotents $\gamma_{\nu,\mathcal{O}}$ and $\gamma_{\nu,K}$.

We will need the following key facts, due originally to I. Gelfand and M. Graev [GG] and S.Gelfand [Ge] (see e.g. [Z, Proposition 9.4]), about the Gelfand-Graev representation in characteristic 0, all of which we deduce here from standard results in [C]:

2.5d. **Theorem.** *For any* $\mu \vDash n$,

(i) ${}^*R^{G_n}_{G_\mu} \Gamma_{n,K} \cong \Gamma_{\mu,K}$;

(ii) $\Gamma_{\mu,K}$ *is a multiplicity-free* KG_μ*-module;*

(iii) *the number of irreducible constituents of* $\Gamma_{n,K}$ *is equal to the number of semisimple conjugacy classes in* G_n*, namely,* $(q-1)q^{n-1}$;

(iv) *for any block-diagonal element* $s \in G_n$*,* $\dim \mathrm{Hom}_{KG_n}(\Gamma_{n,K}, M(s)_K) = 1$;

(v) *given* $\sigma \in \bar{\mathbb{F}}_q^\times$ *of degree* d *over* $\mathbb{F}_q$ *and* $n = kd$*, the image of any non-zero homomorphism from* $\Gamma_{n,K}$ *to* $M^k(\sigma)_K$ *is precisely the irreducible module* $M^k(\sigma)_K y_k$.

Proof. (i) By [C, Theorem 8.1.5], ${}^*R^{G_n}_{G_\mu} \Gamma_{n,K}$ and $\Gamma_{\mu,K}$ have the same character.

(ii) This is [C, Theorem 8.1.3].

(iii) That the number of semisimple classes in G_n is equal to $(q-1)q^{n-1}$ is proved for example in [C, Theorem 3.7.6(ii)]. So now the result follows using (ii) and [C, Proposition 8.3.1].

(iv) Note that by (i) and adjointness,

$$\mathrm{Hom}_{KG_n}(\Gamma_{n,K}, M(s)_K) \cong \mathrm{Hom}_{KG_{\delta(s)}}(\Gamma_{\delta(s),K}, N_K)$$

where N_K is the cuspidal module of (2.4.4). The right hand side is either 0 or 1-dimensional by (ii). To see that it is actually always 1-dimensional, use (iii) and the observation that the sum as s runs over all semisimple classes of G_n of $\dim \mathrm{Hom}_{KG_n}(\Gamma_{n,K}, M(s)_K)$ must count the total number of irreducible constituents of $\Gamma_{n,K}$.

(v) By (iv) and Frobenius reciprocity, the space $\gamma_{n,K} M^k(\sigma)_K$ is one dimensional. So $\gamma_{n,K} M^k(\sigma)_K$ is a one dimensional right module for the Hecke algebra $H_{k,K}$. Since there are just two one dimensional right $H_{k,K}$-modules, namely $\mathcal{I}_{H_{k,K}}$ and $\mathcal{E}_{H_{k,K}}$ (or rather, the analogous right modules), we deduce that $H_{k,K}$ acts on $KG_n\gamma_{n,K} M^k(\sigma)_K$ either by $\mathcal{I}_{H_{k,K}}$ or by $\mathcal{E}_{H_{k,K}}$. So, as x_k and y_k are idempotents up to a scalar over K, the image Y of $\Gamma_{n,K}$ in $M^k(\sigma)_K$ under any non-zero homomorphism satisfies either $Y = M^k(\sigma)_K x_k$ or $Y = M^k(\sigma)_K y_k$. But the dimension of Y is calculated in [C, Theorem 8.4.9] (thanks to [C, Propositions 8.4.4-8.4.5] and (iii)):

$$\dim Y = |G_n : C_{G_n}(s)|_{q'} |C_{G_n}(s)|_q$$

where $s = (\sigma)^k$. Recalling that $C_{G_n}(s) \cong GL_k(\mathbb{F}_{q^d})$, this is easily checked to be the same as dim $M^k(\sigma)_K y_k = \chi_{\sigma,(1^k)}(1)$ as in (2.5.3), not dim $M^k(\sigma)_K x_k = \chi_{\sigma,(k)}(1)$ which is (2.5.2). □

Since Γ_n is a projective FG_n-module, many of the properties in Theorem 2.5d generalize to the modular case:

2.5e. **Corollary.** *For any $\mu \vDash n$ and R equal to F or $\mathcal{O}$,*

(i) ${}^*R^{G_n}_{G_\mu}\Gamma_{n,R} \cong \Gamma_{\mu,R}$;

(ii) *for any block-diagonal $s \in G_n$, $\mathrm{Hom}_{RG_n}(\Gamma_{n,R}, M(s)_R)$ is R-free of rank 1;*

(iii) *given $\sigma \in \bar{\mathbb{F}}_q^\times$ of degree d over $\mathbb{F}_q$ and $n = kd$, we have $mT_w = (-1)^{\ell(w)}m$ for all $w \in \Sigma_k$ and for all m lying in the image of any non-zero homomorphism from $\Gamma_{n,R}$ to $M^k(\sigma)_R$.*

Proof. (i) Since Harish-Chandra restriction commutes with base change, we see easily from Theorem 2.5d(i) that ${}^*R^{G_n}_{G_\mu}\Gamma_n$ and Γ_μ have the same Brauer characters. By (2.2a), ${}^*R^{G_n}_{G_\mu}$ sends projectives to projectives. Therefore ${}^*R^{G_n}_{G_\mu}\Gamma_n$ and Γ_μ are projective FG_μ-modules having the same Brauer character, so they are isomorphic. The result over $\mathcal{O}$ follows immediately since any projective FG_μ-module has a unique lift to $\mathcal{O}$.

(ii) Taking $R = \mathcal{O}$, $\mathrm{Hom}_{\mathcal{O}G_n}(\Gamma_{n,\mathcal{O}}, M(s)_\mathcal{O})$ is an $\mathcal{O}$-lattice in the corresponding space over K, hence is $\mathcal{O}$-free of rank 1 by Theorem 2.5d(iv). By the universal coefficient theorem, $F \otimes_\mathcal{O} \mathrm{Hom}_{\mathcal{O}G_n}(\Gamma_{n,\mathcal{O}}, M(s)_\mathcal{O}) \cong \mathrm{Hom}_{FG_n}(\Gamma_n, M(s))$ as $\Gamma_{n,\mathcal{O}}$ is projective. This implies the result over F.

(iii) Let $\theta_\mathcal{O}$ be a generator of $\mathrm{Hom}_{\mathcal{O}G_n}(\Gamma_{n,\mathcal{O}}, M^k(\sigma)_\mathcal{O})$, applying (ii). Extending scalars, we obtain non-zero maps

$$\theta_K \in \mathrm{Hom}_{KG_n}(\Gamma_{n,K}, M^k(\sigma)_K), \qquad \theta_F \in \mathrm{Hom}_{FG_n}(\Gamma_n, M^k(\sigma)).$$

By Theorem 2.5d(v), the image of θ_K is $M^k(\sigma)_K y_k$, so by (1.1b), for any $v \in \Gamma_{n,K}$, $\theta_K(v)T_w = (-1)^{\ell(w)}\theta_K(v)$ for all $w \in \Sigma_k$. In particular, we see that for $v \in \Gamma_{n,\mathcal{O}} \subset \Gamma_{n,K}$, $\theta_\mathcal{O}(v)T_w = (-1)^{\ell(w)}\theta_\mathcal{O}(v)$ for all $w \in \Sigma_k$. This proves the result in case $R = \mathcal{O}$.

Over F, $\theta_F(\Gamma_n)$ is certainly a quotient of $F \otimes_\mathcal{O} \theta_\mathcal{O}(\Gamma_{n,\mathcal{O}})$ (though we cannot yet assert that the two are isomorphic). Since the action of $H_{k,R}$ on $M^k(\sigma)_R$ is compatible with base change, the conclusion follows directly. □

Finally, we can now prove that $M(\sigma)$ is an irreducible FG_n-module, as we mentioned earlier, at least if σ is a p-regular element. Later, we will also prove this if σ is not p-regular. We remark that the first proof of this fact (for arbitrary σ) was given in [D1, D2] using the classification of irreducible FG_n-modules. There is also a simple direct proof due to James [J2, §3], which depends on the earlier work of Gelfand [Ge].

2.5f. **Lemma.** *If $\sigma \in \bar{\mathbb{F}}_q^\times$ is p-regular of degree n over $\mathbb{F}_q$, the FG_n-module $M(\sigma)$ is irreducible.*

Proof. In view of Lemma 2.4c, we know that there is just one irreducible FG_n-module, N say, in the same block as $M(\sigma)$. In other words, all composition factors of $M(\sigma)$ are isomorphic to N. But by Corollary 2.5e(ii), Γ_n is a projective FG_n-module such that $\mathrm{Hom}_{FG_n}(\Gamma_n, M(\sigma))$ is one dimensional. So, Γ_n must contain the projective cover of N as a summand and now the one dimensionality implies that N appears with multiplicity one as a constituent of $M(\sigma)$, hence $M(\sigma) = N$ is irreducible. □

Chapter 3

Connecting GL_n with quantum linear groups

In this chapter, we prove the Morita theorem at the heart of the modular theory. This was first proved by Cline, Parshall and Scott [CPS$_3$, §9]. The proof in *loc. cit.* depends fundamentally on the work of James [J$_2$] and Dipper-James [DJ$_3$], whereas the approach here is self-contained, independent of this earlier work.

3.1. Schur functors

We begin with a short review of the "Schur functors" used heavily throughout the remainder of the chapter. The results described here are based on the results in [G$_2$, chapter 6] and [JS], though we work in terms of projective modules instead of idempotents. We also mention the work of Auslander [A], where a thorough study of the functors in this section was made in a more general setting.

Fix a finite dimensional F-algebra C (for this section only, F can be taken to be an arbitrary field). Assume that we are given a fixed projective module $P \in \text{mod}(C)$. We set $H = \text{End}_C(P)$, writing endomorphisms commuting with the left C-action on the right. Define the functors

$$\alpha : \text{mod}(C) \to \text{mod}(H), \qquad \alpha = \text{Hom}_C(P, ?), \tag{3.1.1}$$

$$\beta : \text{mod}(H) \to \text{mod}(C), \qquad \beta = P \otimes_H ?. \tag{3.1.2}$$

Since P is projective, α is exact. Moreover, by adjointness of 'tensor' and 'hom', β is left adjoint to α. The basic example to keep in mind (e.g. as in (1.5.2)) is the case that $P = Ce$ for some idempotent $e \in C$, when $H \cong eCe$. Then the functor α is the familiar *Schur functor*, since $\text{Hom}_C(Ce, V) \cong eV$ for any left C-module V, and β is the *inverse Schur functor.*

Given a left C-module V, let $O_P(V)$ denote the largest submodule V' of V with the property that $\text{Hom}_C(P, V') = 0$. Let $O^P(V)$ denote submodule of V generated by the images of all C-homomorphisms from P to V. Since P is projective, these have alternative descriptions: $O_P(V)$ (resp. $O^P(V)$) is the largest (resp. smallest) submodule V' of V such that no composition factor of V' (resp. V/V') appears in the head of P. We remark that in [A], Auslander refers to $O^P(V)$ as the P-trace and $O_P(V)$ as the P-torsion part of V.

Clearly any C-module homomorphism $V \to W$ sends $O_P(V)$ into $O_P(W)$ and $O^P(V)$ into $O^P(W)$, so we can view O_P and O^P as functors $\mathrm{mod}(C) \to \mathrm{mod}(C)$, by defining their action on morphisms to be restriction. Finally, any homomorphism $V \to W$ induces a well-defined C-module homomorphism $V/O_P(V) \to W/O_P(W)$. We thus obtain an exact functor $A_P : \mathrm{mod}(C) \to \mathrm{mod}(C)$ defined on objects by $V \mapsto V/O_P(V)$.

3.1a. **Lemma.** *The composite functors $\alpha \circ \beta$ and $\alpha \circ A_P \circ \beta$ are both isomorphic to the identity.*

Proof. For any left H-module U, the fact that P is projective and [AF, 20.10] implies that

$$\mathrm{Hom}_C(P, P \otimes_H U) \cong \mathrm{Hom}_C(P, P) \otimes_H U = H \otimes_H U \cong U.$$

All isomorphisms are natural, so this proves that $\alpha \circ \beta$ is isomorphic to the identity. Now apply the exact functor α to the exact sequence $0 \to O_P \circ \beta(U) \to \beta(U) \to A_P \circ \beta(U) \to 0$, using the fact that $\mathrm{Hom}_C(P, O_P \circ \beta(U)) = 0$, to deduce that $\alpha \circ \beta(U)$ and $\alpha \circ A_P \circ \beta(U)$ are naturally isomorphic, completing the proof. □

3.1b. **Lemma.** *For $V \in \mathrm{mod}(C)$, let $\hat{V} = \beta \circ \alpha(V) = P \otimes_H \mathrm{Hom}_C(P, V)$ and let $\omega : \hat{V} \to V$ be the natural C-homomorphism defined by $p \otimes \varphi \mapsto \varphi(p)$ for $\varphi \in \mathrm{Hom}_C(P, V)$ and $p \in P$. Then, $\mathrm{im}\,\omega = O^P(V)$ and $\ker \omega \subseteq O_P(\hat{V})$.*

Proof. It follows directly from the definitions that $\mathrm{im}\,\omega = O^P(V)$. Let $Z = \ker \omega$; we need to show that $\alpha(Z) = \mathrm{Hom}_C(P, Z) = 0$. Using the short exact sequence

$$0 \longrightarrow \alpha(Z) \longrightarrow \alpha(\hat{V}) \longrightarrow \alpha(O^P(V)) \longrightarrow 0,$$

we just need to check that $\alpha(\hat{V}) \cong \alpha(O^P(V))$. By Lemma 3.1a, $\alpha(\hat{V}) \cong \mathrm{Hom}_C(P, V)$ which is obviously isomorphic to $\alpha(O^P(V)) = \mathrm{Hom}_C(P, O^P(V))$ by definition of O^P. □

3.1c. **Corollary.** *If $V, W \in \mathrm{mod}(C)$ satisfy $O^P V = V$ and $O_P W = 0$, then*

$$\mathrm{Hom}_C(V, W) \cong \mathrm{Hom}_H(\alpha(V), \alpha(W)).$$

Proof. By adjointness, $\mathrm{Hom}_H(\alpha(V), \alpha(W)) \cong \mathrm{Hom}_C(\beta \circ \alpha(V), W)$. By the lemma, there is a natural homomorphism $\beta \circ \alpha(V) \to V$ which is surjective as $O^P V = V$, and has kernel Z contained in $O_P(\beta \circ \alpha(V))$. Moreover, any homomorphism from $\beta \circ \alpha(V)$ to W must act as zero on Z since $O_P(W) = 0$, hence factors through the quotient V of $\beta \circ \alpha(V)$. Thus, $\mathrm{Hom}_C(\beta \circ \alpha(V), W) \cong \mathrm{Hom}_C(V, W)$. □

Now we have the main result about Schur functors, see [JS, §2]:

3.1d. **Theorem.** *The restrictions of the functors α and $A_P \circ \beta$ induce mutually inverse equivalences of categories between* $\mathrm{mod}(H)$ *and the full subcategory* $\mathfrak{M}$ *of* $\mathrm{mod}(C)$ *consisting of all* $V \in \mathrm{mod}(C)$ *such that* $O_P(V) = 0, O^P(V) = V$.

Proof. We first note that $A_P \circ \beta$ is a well-defined functor from $\mathrm{mod}(H)$ to $\mathfrak{M}$. Take $U \in \mathrm{mod}(H)$. Then, $A_P \circ \beta(U)$ is a quotient of $\beta(U)$ which is a quotient of the left C-module $P \otimes_F U \cong P^{\oplus \dim U}$. The latter is certainly generated by the images of all C-homomorphisms from P, so $O^P(A_P \circ \beta(U)) = A_P \circ \beta(U)$. Moreover, $O_P(A_P \circ \beta(U)) = 0$, so we do indeed have that $A_P \circ \beta(U) \in \mathfrak{M}$.

Now for the theorem, take $V \in \mathfrak{M}$ and consider $A_P \circ \beta \circ \alpha(V)$. By Lemma 3.1b and the assumption that $O^P(V) = V$, we know that $\beta \circ \alpha(V)$ has a submodule $Z \subseteq O_P(\beta \circ \alpha(V))$ such that $(\beta \circ \alpha(V))/Z \cong V$. Since $O_P(V) = 0$, we see that in fact $Z = O_P(\beta \circ \alpha(V))$, so $A_P \circ \beta \circ \alpha(V) \cong V$, and this isomorphism is certainly functorial. Finally, by Lemma 3.1a, we know already that $\alpha \circ A_P \circ \beta$ is isomorphic to the identity, completing the proof. □

3.1e. **Corollary.** *Let $\{E_i \mid i \in I\}$ be a complete set of non-isomorphic irreducible C-modules appearing in the head of P. For $i \in I$, set $D_i = \alpha(E_i)$. Then, the set $\{D_i \mid i \in I\}$ is a complete set of non-isomorphic irreducible H-modules, and $A_P \circ \beta(D_i) \cong E_i$.*

Proof. Note that each E_i ($i \in I$) is an irreducible module belonging to the Abelian category $\mathfrak{M}$. Consequently, by Theorem 3.1d, each D_i is an irreducible H-module with $A_P \circ \beta(D_i) \cong E_i$, and the D_i's are pairwise non-isomorphic. To see that all irreducible H-modules arise in this way, use Fitting's lemma. □

Finally, we include a useful lemma which gives a more explicit description of the effect of the composite functor $A_P \circ \beta$ on left ideals of H:

3.1f. **Lemma.** *Suppose that $O_P(P) = 0$, that is, that every composition factor of the socle of P also appears in its head. Then for any left ideal J of H, $A_P \circ \beta(J) \cong PJ$.*

Proof. Our assumption on P implies that $O_P(PJ) = 0$ hence $A_P(PJ) \cong PJ$. Now we prove the more general result that $A_P \circ \beta(J) \cong A_P(PJ)$, without any assumption on P. There is a short exact sequence $0 \longrightarrow Z \longrightarrow P \otimes_H J \xrightarrow{\mu} PJ \longrightarrow 0$ where μ is the natural multiplication map. Applying the exact functor A_P to this, we see that it suffices to show that $A_P(Z) = 0$, or equivalently, that $\mathrm{Hom}_C(P, Z) = 0$.

Now apply α to this short exact sequence, using Lemma 3.1a, to obtain the exact sequence: $0 \longrightarrow \mathrm{Hom}_C(P, Z) \longrightarrow J \xrightarrow{\bar{\mu}} \mathrm{Hom}_C(P, PJ) \longrightarrow 0$. The second map $\bar{\mu}$ maps $j \in J$ to the homomorphism $P \to PJ$ given by right multiplication by j. Since P is a faithful H-module, $\bar{\mu}$ is injective. This implies that $\mathrm{Hom}_C(P, Z) = 0$ as required. □

3.2. The cuspidal algebra

Now we come to a key technical lemma underlying the modular theory; it will allow us to apply the results of §3.1 to study the module $M^k(\sigma)$ from (2.4.2). This lemma was noticed originally (in a slightly different form) by Cline, Parshall and Scott [CPS3, Lemma 9.1] and simplifies the original theory of [DJ3] considerably. The version we present here is due to V. Schubert [S, 12.4/1]; we are grateful to Schubert for allowing us to include the proof of this lemma.

3.2a. **Lemma.** *Let A be a ring and $0 \longrightarrow Z \longrightarrow P \stackrel{\pi}{\longrightarrow} M \longrightarrow 0$ be a short exact sequence of A-modules with P projective. If every A-module homomorphism from P to M annihilates Z, then M is a projective $A/\operatorname{ann}_A(M)$-module, where* $\operatorname{ann}_A(M)$ *denotes the annihilator of M in A.*

Proof. We need to show that every $A/\operatorname{ann}_A(M)$-module homomorphism from M to a quotient of an $A/\operatorname{ann}_A(M)$-module V can be lifted to a homomorphism to V. Equivalently, we show that every A-module homomorphism $\alpha : M \to U$, where U is a quotient of an A-module V with $\operatorname{ann}_A(M) \subseteq \operatorname{ann}_A(V)$, can be lifted to a homomorphism $\beta : M \to V$. Well, $\alpha \circ \pi$ gives an A-module homomorphism from P to U so can be lifted (as P is projective) to a map $\gamma : P \to V$. So the result will follow if we can show that γ annihilates Z, so that γ factors through $P \stackrel{\pi}{\longrightarrow} M$ to induce the required map β. In other words, we need to check that every A-module homomorphism $\gamma : P \to V$ annihilates Z, for every A-module V with $\operatorname{ann}_A(M) \subseteq \operatorname{ann}_A(V)$.

Given such a module V, the map $\operatorname{Hom}_A(P, A) \otimes_A V \to \operatorname{Hom}_A(P, V)$ sending a generator $f \otimes v \in \operatorname{Hom}_A(P, A) \otimes_A V$ to the map $p \mapsto f(p)v$ for $p \in P$, is an isomorphism by [AF, 20.10] (this requires the projectivity of P). Therefore, we just need to check that $f(p)v = 0$ for all $f \in \operatorname{Hom}_A(P, A)$, $p \in Z$ and $v \in V$, or in other words, that $f(Z) \subseteq \operatorname{ann}_A(V)$ for all $f \in \operatorname{Hom}_A(P, A)$. Now take such an $f \in \operatorname{Hom}_A(P, A)$. For $m \in M$, the map $P \to M$ defined by $p \mapsto f(p)m$ for $p \in P$ is an A-module homomorphism, so by hypothesis annihilates Z. That is, $f(Z) \subseteq \operatorname{ann}_A(M)$. By our initial assumption on V, this finally implies that $f(Z) \subseteq \operatorname{ann}_A(V)$ as required. □

3.2b. **Remark.** In [S], Schubert also proves the converse to this lemma, namely, if M is a projective $A/\operatorname{ann}_A(M)$-module, then every A-module homomorphism from P to M annihilates Z. The latter condition is easily checked to be equivalent to the hypothesis adopted in the earlier work of Dipper [D3, D4].

Now *for the remainder of the chapter*, we fix $\sigma \in \bar{\mathbb{F}}_q^\times$ of degree d over $\mathbb{F}_q$. We assume moreover that the following two properties are satisfied:

(A1) $M(\sigma)$ is an irreducible FG_d-module;

(A2) the FG_{kd}-module $M^k(\sigma)$ has exactly Par_k non-isomorphic composition factors for all $k \geq 1$.

We know for instance by Lemma 2.4c and Lemma 2.5f that the assumptions (A1) and (A2) are satisfied if σ is p-regular. In fact, we will prove later in Theorem 4.3b that (A1) and (A2) are satisfied for *arbitrary* σ, so these assumptions will turn out to be redundant.

Fix also now some $k \geq 1$ and set $n = kd$. As our first consequence of the assumptions on σ, we have:

3.2c. **Lemma.** *The module $M^k(\sigma)$ is self-dual, that is, $M^k(\sigma) \cong M^k(\sigma)^\tau$. In particular, the socle and the head of $M^k(\sigma)$ are isomorphic.*

Proof. Since $M(\sigma)$ is irreducible by assumption, this follows immediately as a special case of Corollary 2.2f. □

Introduce the *cuspidal algebra*

$$C_{\sigma,k} = C_{F,(\sigma)^k}(GL_n(\mathbb{F}_q)) = FG_n/\operatorname{ann}_{FG_n}(M^k(\sigma)). \tag{3.2.1}$$

So, $C_{\sigma,k}$ is the image of FG_n under the representation afforded by the FG_n-module $M^k(\sigma)$. Note in view of (2.4b) that for p-regular σ, $C_{\sigma,k}$ is a quotient algebra of the corresponding block algebra $B_{\sigma,k}$ of FG_n (as defined at the end of §2.4).

3.2d. **Theorem.** *$M^k(\sigma)$ is a projective $C_{\sigma,k}$-module with endomorphism algebra isomorphic to $H_k = H_{F,q^d}(\Sigma_k)$, acting as in (2.5b).*

Proof. Write $G = G_n$, $L = G_{(d^k)}$, $N = M(\sigma)\boxtimes\cdots\boxtimes M(\sigma)$ and $M = M^k(\sigma) \cong R_L^G N$ for short. By our assumption (A1), N is an irreducible cuspidal FL-module. Let Q be the projective cover of N in the category $\operatorname{mod}(FL)$. By the Mackey formula,

$$\begin{aligned}\alpha &:= \dim \operatorname{Hom}_{FG}(R_L^G(Q), M) = \dim \operatorname{Hom}_{FL}(Q, {}^*R_L^G \circ R_L^G(N)) \\ &= \sum_{w \in D_{(d^k),(d^k)}} \dim \operatorname{Hom}_{FL}(Q, R^L_{L\cap {}^wL} \circ \operatorname{conj}_w \circ {}^*R^L_{L\cap {}^{w^{-1}}L}(N)).\end{aligned}$$

Now since N is cuspidal, (2.1c) gives that the summand corresponding to w is zero unless $w = \pi_x$ for some $x \in \Sigma_k$. So, $\alpha = \sum_{x\in\Sigma_k} \dim \operatorname{Hom}_{FL}(Q, \operatorname{conj}_{\pi_x}(N))$. Finally, observe that $\operatorname{conj}_{\pi_x}(N) \cong N$ as an FL-module (the isomorphism sends $v_1 \otimes \cdots \otimes v_k \in N$ to $v_{x1} \otimes \cdots \otimes v_{xk} \in \operatorname{conj}_{\pi_x}(N)$). So since N is irreducible and Q is its projective cover, we see that $\alpha = \sum_{x\in\Sigma_k} \dim \operatorname{Hom}_{FL}(Q, N) = k!$.

As Q is projective and maps surjectively onto N, $R_L^G Q$ is projective and maps surjectively onto M. So, $R_L^G Q$ contains the projective cover P of M as a summand. Also recall from (2.5b) that H_k, which has dimension $k!$, embeds into $\operatorname{End}_{FG}(M)$. Combining these remarks with our calculation that $\alpha = k!$, we have shown that

$$\dim \operatorname{Hom}_{FG}(P, M) \leq \dim \operatorname{Hom}_{FG}(R_L^G Q, M) = k! \leq \dim \operatorname{Hom}_{FG}(M, M),$$

whence that equality holds everywhere. Now the criterion of Lemma 3.2a implies that M is a projective $C_{\sigma,k}$-module, while by dimension, the embedding of H_k into $\operatorname{End}_{FG}(M^k(\sigma))$ from (2.5b) is an isomorphism. □

We mention the following immediate corollary, which is relevant to the point of view (not pursued further here) of modular Harish-Chandra theory:

3.2e. **Corollary.** *There is a bijection between the isomorphism classes of irreducible H_k-modules and the isomorphism classes of irreducible FG_n-modules appearing as constituents of the head (resp. socle) of $M^k(\sigma)$.*

Proof. By Lemma 3.2c, a copy of every composition factor of the socle of $M^k(\sigma)$ appears in its head. So the corollary follows from the theorem by Corollary 3.1e, taking $C = C_{\sigma,k}, H = H_k, P = M^k(\sigma)$. □

Suppose now that $\nu = (k_1, \dots, k_a) \models k$. Regard $M^{k_i}(\sigma)$ as an (FG_{dk_i}, H_{k_i})-bimodule for each $i = 1, \dots, a$ in the same way as explained after (2.5b). Then, identifying $FG_{d\nu}$ with $FG_{dk_1} \otimes \cdots \otimes FG_{dk_a}$ and H_ν with $H_{k_1} \otimes \cdots \otimes H_{k_a}$, we obtain an $(FG_{d\nu}, H_\nu)$-bimodule:

$$M^\nu(\sigma) = M^{k_1}(\sigma) \boxtimes \cdots \boxtimes M^{k_a}(\sigma) \cong R^{G_{d\nu}}_{G_{(d^k)}}\big(\underbrace{M(\sigma) \boxtimes \cdots \boxtimes M(\sigma)}_{k \text{ times}}\big). \tag{3.2.2}$$

We have the Levi analogue of the cuspidal algebra, namely,

$$C_{\sigma,\nu} = FG_{d\nu} / \operatorname{ann}_{FG_{d\nu}}(M^\nu(\sigma)) \cong C_{\sigma,k_1} \otimes \cdots \otimes C_{\sigma,k_a}. \tag{3.2.3}$$

Theorem 3.2d immediately gives that $M^\nu(\sigma)$ is projective as a $C_{\sigma,\nu}$-module, and that H_ν is precisely the endomorphism algebra $\operatorname{End}_{C_{\sigma,\nu}}(M^\nu(\sigma))$.

Now, $M^\nu(\sigma)$ is an $(FG_{d\nu}, H_\nu)$-bimodule, so we can regard $R^{G_n}_{G_{d\nu}} M^\nu(\sigma)$ as an (FG_n, H_ν)-bimodule. Similarly, $M^k(\sigma)$ is an (FG_n, H_k)-bimodule, so ${}^*R^{G_n}_{G_{d\nu}} M^k(\sigma)$ is an $(FG_{d\nu}, H_k)$-bimodule in a precise way. On the other hand, $M^k(\sigma)$ is an (FG_n, H_k)-bimodule, hence also an (FG_n, H_ν)-bimodule, restricting the H_k-action to the subalgebra H_ν of H_k, while $M^\nu(\sigma) \otimes_{H_\nu} H_k$ is an $(FG_{d\nu}, H_k)$-bimodule. The following basic lemma identifies these various bimodule structures:

3.2f. **Lemma.** *For $\nu \models k$,*

(i) *$R^{G_n}_{G_{d\nu}} M^\nu(\sigma)$ is isomorphic to $M^k(\sigma)$ as an (FG_n, H_ν)-bimodule;*

(ii) *${}^*R^{G_n}_{G_{d\nu}} M^k(\sigma)$ is isomorphic to $M^\nu(\sigma) \otimes_{H_\nu} H_k$ as an $(FG_{d\nu}, H_k)$-bimodule.*

In particular, the $FG_{d\nu}$-action on ${}^*R^{G_n}_{G_{d\nu}} M^k(\sigma)$ factors through the quotient $C_{\sigma,\nu}$, so ${}^*R^{G_n}_{G_{d\nu}} M^k(\sigma)$ is a $C_{\sigma,\nu}$-module in the natural way.

Proof. (i) Let $N = M(\sigma) \boxtimes \cdots \boxtimes M(\sigma)$ (k times). As in (3.2.2), $M^\nu(\sigma) \cong R^{G_{d\nu}}_{G_{(d^k)}} N$. So by transitivity of Harish-Chandra induction, we see that $M^k(\sigma) \cong R^{G_n}_{G_{d\nu}} M^\nu(\sigma)$ as FG_n-modules. We need to check that the isomorphism is compatible with the right H_ν-module structures. We have a canonical embedding of N into $M^\nu(\sigma)$ (resp. into $M^k(\sigma)$) as an $FG_{(d^k)}$-module. Take $w \in \Sigma_\nu$. Then the action of B_w (hence $T_w \in H_k$) on $M^k(\sigma)$ is by definition the unique FG_n-endomorphism of $M^k(\sigma)$

whose restriction to N is the map A_w of (2.5.1). Similarly, the action of B_w (hence $T_w \in H_\nu$) on $M^\nu(\sigma)$ is the unique $FG_{d\nu}$-endomorphism of $M^\nu(\sigma)$ whose restriction to N is the map A_w. So the induced action of $T_w \in H_\nu$ on $R^{G_n}_{G_{d\nu}} M^\nu(\sigma)$ has the same restriction to N as the action of $T_w \in H_k$ on $M^k(\sigma)$, identifying $R^{G_n}_{G_{d\nu}} M^\nu(\sigma)$ with $M^k(\sigma)$. This shows that the two actions of T_w coincide.

(ii) Writing $N = M(\sigma) \boxtimes \cdots \boxtimes M(\sigma)$ and $L = G_{(d^k)}$, we have

$$ {}^*R^{G_n}_{G_{d\nu}} \circ R^{G_n}_L(N) \cong \bigoplus_{w \in D_{d\nu,(d^k)}} R^{G_{d\nu}}_{G_{d\nu} \cap {}^w L} \circ \mathrm{conj}_w \circ {}^*R^L_{L \cap {}^{w^{-1}} G_{d\nu}}(N). $$

As N is a cuspidal FL-module the summand corresponding to w is zero unless $L \cap {}^{w^{-1}} G_{d\nu} = L$. In that case, by (2.1c), $w = \pi_x$ for $x \in D_{\nu,(1^k)} = D_\nu^{-1}$. Now recalling that $\mathrm{conj}_{\pi_x}(N) \cong N$ as an FL-module, we have shown that

$$ {}^*R^{G_n}_{G_{d\nu}} M^k(\sigma) \cong \bigoplus_{x \in D_\nu^{-1}} R^{G_{d\nu}}_L(N) \cong M^\nu(\sigma)^{\oplus |D_\nu|} $$

as an $FG_{d\nu}$-module. In particular, this shows that the $FG_{d\nu}$-action on ${}^*R^{G_n}_{G_{d\nu}} M^k(\sigma)$ factors through $C_{\sigma,\nu}$.

By (i), we can identify $M^k(\sigma)$ and $FG_n \otimes_{FP} M^\nu(\sigma)$ as (FG_n, H_ν)-bimodules, where P is the standard parabolic subgroup with Levi factor $G_{d\nu}$. This allows us to identify $M^\nu(\sigma)$ with the $(FG_{d\nu}, H_\nu)$-subbimodule $1 \otimes_{FP} M^\nu(\sigma)$ of $M^k(\sigma)$. Multiplication then gives us an $(FG_{d\nu}, H_k)$-bimodule map $\mu : M^\nu(\sigma) \otimes_{H_\nu} H_k \to M^k(\sigma)$ whose image $M^\nu(\sigma) H_k$ is clearly contained in the fixed point set ${}^*R^{G_n}_{G_{d\nu}} M^k(\sigma)$. Moreover, the calculation in the preceding paragraph shows that

$$ \dim M^\nu(\sigma) \otimes_{H_\nu} H_k = \dim {}^*R^{G_n}_{G_{d\nu}} M^k(\sigma). $$

So it just remains to check that μ is injective.

We have an exact sequence

$$ 0 \longrightarrow Z \longrightarrow M^\nu(\sigma) \otimes_{H_\nu} H_k \xrightarrow{\mu} M^\nu(\sigma) H_k \longrightarrow 0 $$

of $C_{\sigma,\nu}$-modules. Since $M^\nu(\sigma) \otimes_{H_\nu} H_k \cong M^\nu(\sigma)^{\oplus |D_\nu|}$ as an $FG_{d\nu}$-module, and $M^\nu(\sigma)$ is self-dual by Corollary 2.2f, every constituent of the socle of $M^\nu(\sigma) \otimes_{H_\nu} H_k$ appears in the head of $M^\nu(\sigma)$. Therefore to show that $Z = 0$, it suffices to show that $\mathrm{Hom}_{C_{\sigma,\nu}}(M^\nu(\sigma), Z) = 0$. Applying the exact functor $\mathrm{Hom}_{C_{\sigma,\nu}}(M^\nu(\sigma), ?)$ to the above exact sequence using Lemma 3.1a, we obtain the exact sequence

$$ 0 \longrightarrow \mathrm{Hom}_{C_{\sigma,\nu}}(M^\nu(\sigma), Z) \longrightarrow H_k \xrightarrow{\bar{\mu}} \mathrm{Hom}_{C_{\sigma,\nu}}(M^\nu(\sigma), M^\nu(\sigma) H_k) \longrightarrow 0. $$

We just need to check that the map $\bar{\mu}$, which sends $h \in H_k$ to the map $M^\nu(\sigma) \to M^\nu(\sigma) H_k$ given by right multiplication by h, is injective. Suppose for some $h \in H_k$ that $\bar{\mu}(h) = 0$ so $M^\nu(\sigma) h = 0$. Then, as $M^\nu(\sigma)$ generates $M^k(\sigma) = R^{G_n}_{G_{d\nu}} M^\nu(\sigma)$ as an FG_n-module, we deduce that h annihilates all of $M^k(\sigma)$, whence $h = 0$. □

3.2g. **Corollary.** *The following pairs of functors are isomorphic:*
(i) $R_{G_{d\nu}}^{G_n} \circ (M^\nu(\sigma) \otimes_{H_\nu} ?)$ *and* $(M^k(\sigma) \otimes_{H_k} ?) \circ \mathrm{ind}_{H_\nu}^{H_k} : \mathrm{mod}(H_\nu) \to \mathrm{mod}(FG_n)$;
(ii) ${}^*R_{G_{d\nu}}^{G_n} \circ (M^k(\sigma) \otimes_{H_k} ?)$ *and* $(M^\nu(\sigma) \otimes_{H_\nu} ?) \circ \mathrm{res}_{H_\nu}^{H_k} : \mathrm{mod}(H_k) \to \mathrm{mod}(FG_{d\nu})$.

Proof. (i) Take $N \in \mathrm{mod}(H_\nu)$. Let P denote the standard parabolic subgroup of G_n with Levi factor $G_{d\nu}$. Using Lemma 3.2f(i) and associativity of tensor product, we have natural isomorphisms

$$\begin{aligned} M^k(\sigma) \otimes_{H_k} \mathrm{ind}_{H_\nu}^{H_k} N &= M^k(\sigma) \otimes_{H_k} (H_k \otimes_{H_\nu} N) \cong (M^k(\sigma) \otimes_{H_k} H_k) \otimes_{H_\nu} N \\ &\cong M^k(\sigma) \otimes_{H_\nu} N \cong (FG_n \otimes_{FP} M^\nu(\sigma)) \otimes_{H_\nu} N \\ &\cong FG_n \otimes_{FP} (M^\nu(\sigma) \otimes_{H_\nu} N) = R_{G_{d\nu}}^{G_n} (M^\nu(\sigma) \otimes_{H_\nu} N), \end{aligned}$$

as required.

(ii) We first claim that the functors ${}^*R_{G_{d\nu}}^{G_n} \circ (M^k(\sigma) \otimes_{H_k} ?)$ and $({}^*R_{G_{d\nu}}^{G_n} M^k(\sigma)) \otimes_{H_k} ?$ are isomorphic. Recall that ${}^*R_{G_{d\nu}}^{G_n}$ is defined by taking $Y_{d\nu}$-fixed points, so can be viewed as the functor $\mathrm{Hom}_{FY_{d\nu}}(\mathcal{I}, ?)$ where $\mathcal{I}$ is the trivial representation of $Y_{d\nu}$. Now, $FY_{d\nu}$ is a semisimple algebra so $\mathcal{I}$ is a projective $FY_{d\nu}$-module. So e.g. [AF, 20.10] immediately gives a natural isomorphism

$$\mathrm{Hom}_{FY_{d\nu}}(\mathcal{I}, M^k(\sigma) \otimes_{H_k} N) \cong \mathrm{Hom}_{FY_{d\nu}}(\mathcal{I}, M^k(\sigma)) \otimes_{H_k} N$$

for any H_k-module N, to prove the claim. Hence, using Lemma 3.2f(ii) as well, there are natural isomorphisms

$$\begin{aligned} {}^*R_{G_{d\nu}}^{G_n} (M^k(\sigma) \otimes_{H_k} N) &\cong ({}^*R_{G_{d\nu}}^{G_n} M^k(\sigma)) \otimes_{H_k} N \cong (M^\nu(\sigma) \otimes_{H_\nu} H_k) \otimes_{H_k} N \\ &\cong M^\nu(\sigma) \otimes_{H_\nu} (H_k \otimes_{H_k} N) \cong M^\nu(\sigma) \otimes_{H_\nu} \mathrm{res}_{H_\nu}^{H_k} N \end{aligned}$$

for any $N \in \mathrm{mod}(H_k)$. This completes the proof. □

3.3. 'Symmetric' and 'exterior' powers

Now we are ready to introduce modules which play the role of symmetric and exterior powers in the non-defining characteristic theory. As motivation, recall the defining characteristic theory, where V is a finite dimensional vector space over F and Σ_k acts on the right on $V^{\otimes k}$ by permuting tensors. The *symmetric power* $S^k(V)$ can be defined as the largest quotient of $V^{\otimes k}$ on which Σ_k acts trivially. The dual notion, the *divided power* $Z^k(V)$, is the largest submodule of $V^{\otimes k}$ on which Σ_k acts trivially; in positive characteristic, $S^k(V)$ and $Z^k(V)$ need not be isomorphic as left $GL(V)$-modules. We define the analogues of symmetric and divided powers in our theory by

$$\begin{aligned} S^k(\sigma) &= M^k(\sigma)/\{mh - \mathcal{E}_{H_k}(h)m \mid h \in H_k, m \in M^k(\sigma)\}, \\ Z^k(\sigma) &= \{m \in M^k(\sigma) \mid mh = \mathcal{E}_{H_k}(h)m \text{ for all } h \in H_k\}. \end{aligned}$$

Observe these are left FG_n-modules which factor through the quotient $C_{\sigma,k}$ to induce well-defined $C_{\sigma,k}$-modules.

To define the exterior power in defining characteristic, the definition as the largest quotient (resp. submodule) of $V^{\otimes r}$ on which Σ_k acts as sign is of course wrong in characteristic 2, so the best definition from the point of view of the symmetric group is $\Lambda^k(V) = V^{\otimes k}(\sum_{w \in \Sigma_k}(-1)^{\ell(w)}w)$. We define our analogue, a $C_{\sigma,k}$-module again, by

$$\Lambda^k(\sigma) = M^k(\sigma)x_k.$$

It is perhaps unfortunate that our "symmetric powers" correspond now to the sign representation and our "exterior power" corresponds to the trivial representation, unlike in the classical case.

As an example, in the special case $\sigma = 1$ when $k = n$, the module $\Lambda^n(1)$ is precisely the trivial FG_n-module, whereas $Z^n(1)$ is a modular reduction of the Steinberg module (these statements follow as special cases of Theorem 3.5e and Theorem 4.1c).

3.3a. **Lemma.** *$\Lambda^k(\sigma)$ is an irreducible FG_n-module.*

Proof. We know that $M^k(\sigma)$ is a projective $C_{\sigma,k}$-module and that every composition factor of the socle of $M^k(\sigma)$ appears in its head, by Theorem 3.2d and Lemma 3.2c respectively. Also, the left ideal $H_k x_k$ is an irreducible H_k-module. Using these remarks, the lemma follows at once from the general theory of Schur functors; see Lemma 3.1f and Corollary 3.1e. □

The structure of $S^k(\sigma)$ and $Z^k(\sigma)$ is more subtle. Observe though that by definition and (1.1b), $M^k(\sigma)y_k$ is a quotient of $S^k(\sigma)$ and $Z^k(\sigma)$ contains $M^k(\sigma)y_k$ as a submodule.

3.3b. **Lemma.** (i) *$S^k(\sigma) \cong M^k(\sigma) \otimes_{H_k} \mathcal{E}_{H_k}$;*

(ii) *$S^k(\sigma)$ has simple head equal to the irreducible quotient $M^k(\sigma)y_k$ of $M^k(\sigma)$, and no other composition factors of $S^k(\sigma)$ are isomorphic to quotients of $M^k(\sigma)$;*

(iii) *$Z^k(\sigma) \cong S^k(\sigma)^\tau$;*

(iv) *$Z^k(\sigma)$ has simple socle equal to the irreducible submodule $M^k(\sigma)y_k$ of $M^k(\sigma)$, and no other composition factors of $Z^k(\sigma)$ are isomorphic to submodules of $M^k(\sigma)$.*

Proof. (i) By definition, $M^k(\sigma) \otimes_{H_k} \mathcal{E}_{H_k}$ is the quotient of $M^k(\sigma) \otimes_F \mathcal{E}_{H_k}$ by

$$\{mh \otimes 1 - \mathcal{E}_{H_k}(h)m \otimes 1 \mid h \in H_k, m \in M^k(\sigma)\}.$$

If we identify $M^k(\sigma) \otimes_F \mathcal{E}_{H_k}$ and $M^k(\sigma)$ as vector spaces in the natural way, this immediately gives (i).

(ii) Let α, β and A_P be the functors defined in §3.1, taking the projective module P to be the $C_{\sigma,k}$-module $M^k(\sigma)$ of Theorem 3.2d. Then, (i) shows that $S^k(\sigma) \cong \beta(\mathcal{E}_{H_k})$. Just as in Lemma 3.3a, Lemma 3.1f shows that

$$A_P \circ \beta(\mathcal{E}_{H_k}) \cong A_P(S^k(\sigma)) \cong M^k(\sigma)y_k$$

is an irreducible $C_{\sigma,k}$-module. Hence, $M^k(\sigma)y_k$ appears in the head of $S^k(\sigma)$ and no other composition factors of $S^k(\sigma)$ appear in the head of $M^k(\sigma)$. Since $S^k(\sigma)$ is a quotient of $M^k(\sigma)$, this means that $S^k(\sigma)$ actually has simple head.

(iii) Write $M = M^k(\sigma)$ for short. We can by Corollary 2.2f choose some isomorphism $i : M \to M^\tau$ as FG_n-modules. This choice induces an isomorphism $j : \mathrm{End}_{FG_n}(M) \to \mathrm{End}_{FG_n}(M^\tau)$ with $fj(\theta) = i((i^{-1}f)\theta)$ for all $f \in M^\tau, \theta \in \mathrm{End}_{FG_n}(M)$ (recall we are writing endomorphisms on the right). On the other hand, there is a natural anti-isomorphism $\# : \mathrm{End}_{FG_n}(M) \to \mathrm{End}_{FG_n}(M^\tau)$ defined simply by letting $\theta^\#$ be the dual map to $\theta \in \mathrm{End}_{FG_n}(M)$, that is, $(f\theta^\#)(m) = f(m\theta)$ for all $m \in M, f \in M^\tau = M^*$. Now if we set $\tau = j^{-1} \circ \#$, we have defined an anti-automorphism of $H_k = \mathrm{End}_{FG_n}(M)$. Define a non-degenerate bilinear form on M by $(m, n) = i(m)(n)$ for $m, n \in M$. For any $h \in H_k$ we have

$$(m\tau(h), n) = (i^{-1}(i(m)h^\#), n) = (i(m)h^\#)(n) = i(m)(nh) = (m, nh).$$

In other words, the bilinear form $(.,.)$ is 'contravariant' for the action of H_k with respect to the anti-automorphism τ.

By definition, $S^k(\sigma) = M/J$ where $J = \{mh - \mathcal{E}_{H_k}(h)m \mid h \in H_k, m \in M\}$. So:

$$\begin{aligned} S^k(\sigma)^\tau \cong J^\circ &= \{n \in M \mid (n, mh - \mathcal{E}_{H_k}(h)m) = 0 \text{ for all } m \in M, h \in H_k\} \\ &= \{n \in M \mid (n\tau(h) - \mathcal{E}_{H_k}(h)n, m) = 0 \text{ for all } m \in M, h \in H_k\} \\ &= \{n \in M \mid nh = \mathcal{E}_{H_k}(\tau(h))n \text{ for all } h \in H_k\}. \end{aligned}$$

It now just remains to show that $\mathcal{E}_{H_k}(\tau(h)) = \mathcal{E}_{H_k}(h)$ for all $h \in H_k$.

Certainly $\mathcal{E}_{H_k} \circ \tau$ is a linear representation of H_k, so since the only one dimensional H_k-modules are $\mathcal{E}_{H_k}$ and $\mathcal{I}_{H_k}$, we either have that $\mathcal{E}_{H_k} \circ \tau = \mathcal{E}_{H_k}$ as required, or that $\mathcal{E}_{H_k} \circ \tau = \mathcal{I}_{H_k}$. In the latter case, we see from (1.1b) that $S^k(\sigma)^\tau$ contains Mx_k as an irreducible submodule, whence that $S^k(\sigma)$ contains Mx_k in its head. But this is not so according to (ii) unless in fact $Mx_k \cong My_k$, in which case applying Corollary 3.1e, $\mathcal{I}_{H_k} \cong \mathcal{E}_{H_k}$ and we are done.

(iv) This follows from (ii) on dualizing, using (iii). □

Slightly more generally, for $\nu = (k_1, \dots, k_a) \vDash k$, we have analogous left $C_{\sigma,\nu}$-modules $S^\nu(\sigma)$, $Z^\nu(\sigma)$ and $\Lambda^\nu(\sigma)$:

$$S^\nu(\sigma) = M^\nu(\sigma)/\{mh - \mathcal{E}_{H_\nu}(h)m \mid h \in H_\nu, m \in M^\nu(\sigma)\}, \tag{3.3.1}$$

$$Z^\nu(\sigma) = \{m \in M^\nu(\sigma) \mid mh = \mathcal{E}_{H_\nu}(h)m \text{ for all } h \in H_\nu\}, \tag{3.3.2}$$

$$\Lambda^\nu(\sigma) = M^\nu(\sigma)x_\nu. \tag{3.3.3}$$

Recalling (3.2.2) and (3.2.3), if $\nu = (k_1, \dots, k_a)$ then $S^\nu(\sigma) \cong S^{k_1}(\sigma) \boxtimes \cdots \boxtimes S^{k_a}(\sigma)$, and similarly for Z, Λ. In view of this observation, the basic properties of $S^\nu(\sigma)$, $Z^\nu(\sigma)$ and $\Lambda^\nu(\sigma)$ follow directly from Lemma 3.3a and Lemma 3.3b.

3.3c. **Lemma.** *For any* $\nu \vDash k$, ${}^*R^{G_n}_{G_{d\nu}} S^k(\sigma) \cong S^\nu(\sigma)$ *and* ${}^*R^{G_n}_{G_{d\nu}} Z^k(\sigma) \cong Z^\nu(\sigma)$.

Proof. We just need to prove the first statement, the second following directly from Lemma 3.3b(iii) since Harish-Chandra restriction commutes with contravariant duality. For the first, by Lemma 3.3b(i), $S^k(\sigma) \cong M^k(\sigma) \otimes_{H_k} \mathcal{E}_{H_k}$. So using Corollary 3.2g(ii), we have immediately that

$$^*R^{G_n}_{G_{d\nu}} S^k(\sigma) \cong M^\nu(\sigma) \otimes_{H_\nu} \operatorname{res}^{H_k}_{H_\nu} \mathcal{E}_{H_k} \cong M^\nu(\sigma) \otimes_{H_\nu} \mathcal{E}_{H_\nu}.$$

So by Lemma 3.3b(i) (or rather its Levi analogue) we see that $^*R^{G_n}_{G_{d\nu}} S^k(\sigma) \cong S^\nu(\sigma)$. □

We are mainly interested in what follows in the $C_{\sigma,k}$-modules obtained from $S^\nu(\sigma)$, $Z^\nu(\sigma)$ and $\Lambda^\nu(\sigma)$ by Harish-Chandra induction. Set

$$\dot{S}^\nu(\sigma) = M^k(\sigma)/\{mh - \mathcal{E}_{H_\nu}(h)m \mid h \in H_\nu, m \in M^k(\sigma)\}, \tag{3.3.4}$$

$$\dot{Z}^\nu(\sigma) = \{m \in M^k(\sigma) \mid mh = \mathcal{E}_{H_\nu}(h)m \text{ for all } h \in H_\nu\}, \tag{3.3.5}$$

$$\dot{\Lambda}^\nu(\sigma) = M^k(\sigma)x_\nu. \tag{3.3.6}$$

If we identify $M^k(\sigma)$ with $R^{G_n}_{G_{d\nu}} M^\nu(\sigma)$ as (FG_n, H_ν)-bimodules as in Lemma 3.2f(i), it is easy to check that the quotient $\dot{S}^\nu(\sigma)$ of $M^k(\sigma)$ is identified with the quotient $R^{G_n}_{G_{d\nu}} S^\nu(\sigma)$ of $R^{G_n}_{G_{d\nu}} M^\nu(\sigma)$. Similarly, we can identify $R^{G_n}_{G_{d\nu}} Z^\nu(\sigma)$ with $\dot{Z}^\nu(\sigma)$ and $R^{G_n}_{G_{d\nu}} \Lambda^\nu(\sigma)$ with $\dot{\Lambda}^\nu(\sigma)$.

Note by (1.1b) again that $\dot{Z}^\nu(\sigma)$ contains $M^k(\sigma)y_\nu$ as a submodule. Recall the definitions of the H_k-modules $M^\nu = H_k x_\nu$ and $N^\nu = H_k y_\nu$ from §1.1. The next result generalizes Lemma 3.3b:

3.3d. **Lemma.** *For any $\nu \vDash k$, we have:*

(i) $\dot{S}^\nu(\sigma) \cong M^k(\sigma) \otimes_{H_k} N^\nu$.

(ii) $\dot{Z}^\nu(\sigma) \cong \dot{S}^\nu(\sigma)^\tau$.

(iii) *No composition factors of $\dot{Z}^\nu(\sigma)/M^k(\sigma)y_\nu$ are isomorphic to submodules of $M^k(\sigma)$.*

Proof. (i) Since $\dot{S}^\nu(\sigma) \cong R^{G_n}_{G_{d\nu}} S^\nu(\sigma)$, this is immediate from Corollary 3.2g(i) and the Levi analogue of Lemma 3.3b(i).

(ii) This follows from the Levi analogue of Lemma 3.3b(iii), since Harish-Chandra induction commutes with contravariant duality.

(iii) Let α, β and A_P be the functors defined in §3.1 taking the projective module P to be the $C_{\sigma,k}$-module $M^k(\sigma)$ of Theorem 3.2d. Now, N^ν is the left ideal $H_k y_\nu$ of H_k. So, using Lemma 3.2c and Lemma 3.1f, we know that

$$A_P \circ \beta(N^\nu) \cong M^k(\sigma)y_\nu.$$

So using (i) and the definition of the functor $A_P \circ \beta$, we see that $\dot{S}^\nu(\sigma) \cong \beta(N^\nu)$ is an extension of $M^k(\sigma)y_\nu$ and a module having no composition factors in common with the head (or equivalently by Lemma 3.2c the socle) of $M^k(\sigma)$. Now (iii) follows on dualizing using (ii). □

3.4. Endomorphism algebras

Now fix in addition an integer $h \geq 1$ and let $S_{h,k}$ denote the q^d-Schur algebra $S_{F,q^d}(h,k)$ of §1.2. Our first connection between G_n and the q^d-Schur algebra arises as follows (cf. [DJ$_3$, Theorem 2.24(iii)]):

3.4a. **Theorem.** *There is an algebra isomorphism*

$$S_{h,k} \xrightarrow{\sim} \mathrm{End}_{C_{\sigma,k}}\left(\bigoplus_{\nu\in\Lambda(h,k)} \dot{\Lambda}^\nu(\sigma)\right)$$

under which the natural basis element $\phi^u_{\mu,\lambda}$ of $S_{h,k}$ maps to the endomorphism which is zero on the summand $\dot{\Lambda}^\nu(\sigma)$ for $\nu \neq \mu$ and sends $\dot{\Lambda}^\mu(\sigma)$ into $\dot{\Lambda}^\lambda(\sigma)$ via the homomorphism induced by right multiplication in $M^k(\sigma)$ by $\sum_{w\in\Sigma_\mu u\Sigma_\lambda\cap D_\mu^{-1}} T_w$.

Proof. Let $A_P \circ \beta$ denote the equivalence of categories from Theorem 3.1d, for the projective $C_{\sigma,k}$-module $P = M^k(\sigma)$. According to Lemma 3.1f and Lemma 3.2c,

$$\bigoplus_{\nu\in\Lambda(h,k)} \dot{\Lambda}^\nu(\sigma) \cong A_P \circ \beta\left(\bigoplus_{\nu\in\Lambda(h,k)} M^\nu\right).$$

We deduce that the endomorphism algebras of $\bigoplus_{\nu\in\Lambda(h,k)} \dot{\Lambda}^\nu(\sigma)$ and $\bigoplus_{\nu\in\Lambda(h,k)} M^\nu$ are isomorphic; the latter is $S_{h,k}$ by definition. It remains to check that the image of $\phi^u_{\mu,\lambda}$ under the functor $A_P \circ \beta$ is precisely the endomorphism described, which follows using Lemma 3.1f. □

Recalling from (1.2c) that $S_{h,k}$ can also be described as the endomorphism algebra $\mathrm{End}_{H_k}\left(\bigoplus_{\lambda\in\Lambda(h,k)} N^\lambda\right)$, the same argument as the proof of Theorem 3.4a shows (cf. [DJ$_3$, Theorem 2.24(iv)]):

(3.4b) *There is an algebra isomorphism $S_{h,k} \xrightarrow{\sim} \mathrm{End}_{C_{\sigma,k}}\left(\bigoplus_{\nu\in\Lambda(h,k)} M^k(\sigma)y_\nu\right)$ under which the natural basis element $\phi^u_{\mu,\lambda}$ of $S_{h,k}$ maps to the endomorphism which is zero on the summand $M^k(\sigma)y_\nu$ for $\nu \neq \mu$ and sends $M^k(\sigma)y_\mu$ into $M^k(\sigma)y_\lambda$ via the homomorphism induced by right multiplication in $M^k(\sigma)$ by $\sum_{w\in\Sigma_\mu u\Sigma_\lambda\cap D_\mu^{-1}} T_w^{\#}$.*

The next theorem gives the second important connection between G_n and the q^d-Schur algebra (cf. [DJ$_3$, Theorem 2.24(vi)]):

3.4c. **Theorem.** *There is an algebra isomorphism*

$$S_{h,k} \xrightarrow{\sim} \mathrm{End}_{C_{\sigma,k}}\left(\bigoplus_{\nu\in\Lambda(h,k)} \dot{Z}^\nu(\sigma)\right)$$

under which the natural basis element $\phi^u_{\mu,\lambda}$ of $S_{h,k}$ maps to the endomorphism which is zero on the summand $\dot{Z}^\nu(\sigma)$ for $\nu \neq \mu$ and sends $\dot{Z}^\mu(\sigma)$ into $\dot{Z}^\lambda(\sigma)$ via the homomorphism induced by right multiplication in $M^k(\sigma)$ by $\sum_{w\in\Sigma_\mu u\Sigma_\lambda\cap D_\mu^{-1}} T_w^{\#}$.

Proof. Let us first check that the endomorphisms in the statement are well-defined. We need to observe that as submodules of $M^k(\sigma)$, $\dot{Z}^\mu(\sigma)h \subseteq \dot{Z}^\lambda(\sigma)$ where

$$h = \sum_{w \in \Sigma_\mu u \Sigma_\lambda \cap D_\mu^{-1}} T_w^{\#}.$$

To prove this, it suffices by (3.3.5) to show that $T_s - 1$ annihilates $\dot{Z}^\mu(\sigma)h$ for all basic transpositions $s \in \Sigma_\lambda$. Right multiplication by $h(T_s - 1)$ gives an FG_n-module homomorphism from $\dot{Z}^\mu(\sigma)$ to $M^k(\sigma)$. Obviously, $h(T_s - 1)$ annihilates the submodule $M^k(\sigma)y_\mu$ of $\dot{Z}^\mu(\sigma)$ by (1.1b) and (1.2.2). So in fact, $h(T_s - 1)$ must annihilate all of $\dot{Z}^\mu(\sigma)$ by Lemma 3.3d(iii).

Now let S be the subalgebra of $\operatorname{End}_{FG_n}\left(\bigoplus_{\nu \in \Lambda(h,k)} \dot{Z}^\nu(\sigma)\right)$ consisting of all endomorphisms which stabilize the subspace $\bigoplus_{\nu \in \Lambda(h,k)} M_k(\sigma)y_\nu$. Restriction gives an algebra homomorphism

$$S \to \operatorname{End}_{FG_n}\left(\bigoplus_{\nu \in \Lambda(h,k)} M^k(\sigma)y_\nu\right)$$

which is injective by Lemma 3.3d(iii) and surjective by the previous paragraph and (3.4b). This shows in particular that the endomorphisms of $\bigoplus_{\nu \in \Lambda(h,k)} \dot{Z}^\nu(\sigma)$ defined in the statement of the theorem are linearly independent and span S. It remains to check using dimension that S equals all of $\operatorname{End}_{FG_n}\left(\bigoplus_{\nu \in \Lambda(h,k)} \dot{Z}^\nu(\sigma)\right)$. On expanding the direct sums, this will follow if we can show that

$$\dim \operatorname{Hom}_{FG_n}(\dot{Z}^\mu(\sigma), \dot{Z}^\lambda(\sigma)) = \dim \operatorname{Hom}_{H_k}(N^\lambda, N^\mu)$$

for all $\lambda, \mu \in \Lambda(h,k)$. Let α and β be the functors defined in (3.1.1) and (3.1.2) for $P = M^k(\sigma)$. We calculate using Lemma 3.3d and Lemma 3.1a:

$$\begin{aligned}\operatorname{Hom}_{FG_n}(\dot{Z}^\mu(\sigma), \dot{Z}^\lambda(\sigma)) &\cong \operatorname{Hom}_{FG_n}(\dot{S}^\lambda(\sigma), \dot{S}^\mu(\sigma)) \\ &\cong \operatorname{Hom}_{FG_n}(\beta(N^\lambda), \beta(N^\mu)) \\ &\cong \operatorname{Hom}_{H_k}(N^\lambda, \alpha \circ \beta(N^\mu)) \cong \operatorname{Hom}_{H_k}(N^\lambda, N^\mu).\end{aligned}$$

The result follows. □

To proceed further, we need to utilize the properties of the Gelfand-Graev representation from §2.5. Recall the idempotent $\gamma_n \in FG_n$ from (2.5.5) and its Levi analogue $\gamma_{d\nu} \in FG_{d\nu}$ for $\nu = (k_1, \dots, k_a) \vDash k$. Define

$$\begin{aligned}Y^k(\sigma) &= FG_n\gamma_n M^k(\sigma), \\ Y^\nu(\sigma) &= FG_{d\nu}\gamma_{d\nu} M^\nu(\sigma).\end{aligned}$$

Note these definitions are only temporary: at the end of the section, we will see that $Y^k(\sigma) = Z^k(\sigma)$. It is obvious that $Y^\nu(\sigma) \cong Y^{k_1}(\sigma) \boxtimes \dots \boxtimes Y^{k_a}(\sigma)$, so this will also give that $Y^\nu(\sigma) = Z^\nu(\sigma)$.

Recall for the next lemma that by definition, $Z^k(\sigma)$ (resp. $Z^\nu(\sigma)$) is a submodule of $M^k(\sigma)$ (resp. $M^\nu(\sigma)$).

3.4d. **Lemma.** *For $\nu \models k$,*

(i) *$Y^\nu(\sigma)$ is the image of any non-zero element of the one dimensional space $\mathrm{Hom}_{FG_{d\nu}}(\Gamma_{d\nu}, Z^\nu(\sigma))$;*

(ii) *${}^*R^{G_n}_{G_{d\nu}} Y^k(\sigma) \cong Y^\nu(\sigma)$.*

Proof. (i) According to Corollary 2.5e, the space $\mathrm{Hom}_{FG_n}(\Gamma_n, M^k(\sigma))$ is one dimensional and the image of a non-zero such homomorphism lies in $Z^k(\sigma)$. So, by Frobenius reciprocity and the fact that γ_n is idempotent, $Y^k(\sigma)$ is precisely the image of any non-zero element of the one dimensional space $\mathrm{Hom}_{FG_n}(\Gamma_n, Z^k(\sigma))$. Generalizing to the Levi analogue in the obvious way gives (i).

(ii) We first observe that ${}^*R^{G_n}_{G_{d\nu}} Y^k(\sigma)$ is non-zero. For this, note by (i) that $Y^k(\sigma)$ is non-zero and a submodule of $M^k(\sigma)$, so $\mathrm{Hom}_{FG_n}(Y^k(\sigma), M^k(\sigma)) \neq 0$. Since $\mathrm{Hom}_{FG_{d\nu}}({}^*R^{G_n}_{G_{d\nu}} Y^k(\sigma), M^\nu(\sigma)) \cong \mathrm{Hom}_{FG_n}(Y^k(\sigma), M^k(\sigma))$, ${}^*R^{G_n}_{G_{d\nu}} Y^k(\sigma)$ must therefore also be non-zero.

Now let $\theta : \Gamma_n \to Z^k(\sigma)$ be a non-zero homomorphism, with image $Y^k(\sigma)$ by (i). Applying the exact functor ${}^*R^{G_n}_{G_{d\nu}}$, recalling that ${}^*R^{G_n}_{G_{d\nu}} \Gamma_n \cong \Gamma_{d\nu}$ and ${}^*R^{G_n}_{G_{d\nu}} Z^k(\sigma) \cong Z^\nu(\sigma)$ by Corollary 2.5e(i) and Lemma 3.3c, we obtain a homomorphism $\bar\theta : \Gamma_{d\nu} \to Z^\nu(\sigma)$ with image ${}^*R^{G_n}_{G_{d\nu}} Y^k(\sigma)$. But $\bar\theta$ is non-zero, and by (i) again, the image of such a non-zero homomorphism is precisely $Y^\nu(\sigma)$. □

3.4e. **Lemma.** *Suppose that we are given $\lambda, \mu \models k$ and an $FG_{d\mu}$-submodule M of $M^\mu(\sigma)$. If $w \in D_{d\lambda, d\mu}$ is such that ${}^*R^{G_{d\mu}}_{G_{d\mu} \cap {}^{w^{-1}}G_{d\lambda}}(M) \neq 0$, then w is of the form π_x for some $x \in D_{\lambda,\mu}$.*

Proof. By exactness of Harish-Chandra restriction, it suffices to prove this in the special case $M = M^\mu(\sigma)$. As usual, write $L = G_{(d^k)}$ and $N = M(\sigma) \boxtimes \cdots \boxtimes M(\sigma)$. Define $\nu \models n$ by $G_\nu = G_{d\mu} \cap {}^{w^{-1}}G_{d\lambda}$. Then a Mackey calculation gives that

$$ {}^*R^{G_{d\mu}}_{G_\nu} M \cong \bigoplus_{y \in D^{d\mu}_{\nu,(d^k)}} R^{G_\nu}_{G_\nu \cap {}^yL} \circ \mathrm{conj}_y \circ {}^*R^L_{L \cap {}^{y^{-1}}G_\nu}(N). $$

Since this is non-zero and N is cuspidal, there must be some $y \in D^{d\mu}_{\nu,(d^k)}$ such that $L = L \cap {}^{y^{-1}}G_\nu$, in which case ${}^yL = G_\nu \cap {}^yL$ is a standard Levi subgroup of G_n conjugate to L. But the only such Levi subgroup is L itself, so in fact ${}^yL = L$ and we have that $L \subseteq G_\nu = G_{d\mu} \cap {}^{w^{-1}}G_{d\lambda}$. We deduce from (2.1c) that $w^{-1} = \pi_x$ for some $x \in D_{\mu,\lambda}$. □

The next Mackey calculation is of central importance.

3.4f. **Lemma.** *For any $\lambda, \mu \models k$, all of the spaces*

$$ \mathrm{Hom}_{FG_n}(R^{G_n}_{G_{d\lambda}} Y^\lambda(\sigma), R^{G_n}_{G_{d\mu}} Y^\mu(\sigma)), \tag{3.4.1} $$

$$ \mathrm{Hom}_{FG_n}(R^{G_n}_{G_{d\lambda}} \Gamma_{d\lambda}, R^{G_n}_{G_{d\mu}} Y^\mu(\sigma)), \tag{3.4.2} $$

$$ \mathrm{Hom}_{FG_n}(R^{G_n}_{G_{d\lambda}} \Gamma_{d\lambda}, R^{G_n}_{G_{d\mu}} Z^\mu(\sigma)) \tag{3.4.3} $$

have the same dimension, namely, $|D_{\lambda,\mu}|$.

Proof. Consider (3.4.1). By the Mackey theorem and Lemma 3.4e, we have that

$$\mathrm{Hom}_{FG_n}(R^{G_n}_{G_{d\lambda}} Y^\lambda(\sigma), R^{G_n}_{G_{d\mu}} Y^\mu(\sigma)) \cong \bigoplus_{w \in D_{d\lambda,d\mu}} H_w \cong \bigoplus_{x \in D_{\lambda,\mu}} H_{\pi_x},$$

where for $w \in \Sigma_n$,

$$H_w = \mathrm{Hom}_{FG_{d\lambda}}(Y^\lambda(\sigma), R^{G_{d\lambda}}_{G_{d\lambda}\cap {}^w G_{d\mu}} \circ \mathrm{conj}_w \circ {}^*R^{G_{d\mu}}_{G_{d\mu}\cap {}^{w^{-1}} G_{d\lambda}}(Y^\mu(\sigma))).$$

So to prove the result for (3.4.1), we need to show that each H_{π_x} is one dimensional. Fix $x \in D_{\lambda,\mu}$. Applying (2.1c), $G_{d\lambda} \cap {}^{\pi_x} G_{d\mu} = G_{d\nu}$ and $G_{d\mu} \cap {}^{\pi_x^{-1}} G_{d\lambda} = G_{d\nu'}$ for some $\nu, \nu' \vDash k$. Using Lemma 3.4d(ii) and adjointness,

$$\dim H_{\pi_x} = \dim \mathrm{Hom}_{FG_{d\nu}}(Y^\nu(\sigma), \mathrm{conj}_{\pi_x} Y^{\nu'}(\sigma)).$$

Conjugating the standard Levi subgroup $G_{d\nu'}$ by π_x simply rearranges the diagonal blocks to obtain $G_{d\nu}$. Since $Y^{\nu'}(\sigma)$ is an outer tensor product over these diagonal blocks, we see that $\mathrm{conj}_{\pi_x} Y^{\nu'}(\sigma) \cong Y^\nu(\sigma)$. Finally, $Y^\nu(\sigma)$ is a quotient of $\Gamma_{d\nu}$ and a submodule of $Z^\nu(\sigma)$ so Lemma 3.4d(i) implies that $\dim H_{\pi_x} = \dim \mathrm{Hom}_{FG_{d\nu}}(Y^\nu(\sigma), Y^\nu(\sigma))$ is one dimensional.

This proves the lemma for (3.4.1), and the proofs for (3.4.2) and (3.4.3) are entirely similar, using instead that ${}^*R^{G_n}_{G_{d\lambda}} Z^k(\sigma) \cong Z^\lambda(\sigma)$ and ${}^*R^{G_n}_{G_{d\lambda}} \Gamma_n \cong \Gamma_{d\lambda}$ according to Lemma 3.3c and Corollary 2.5e(i) respectively. □

Now we can prove what we regard as the fundamental theorem:

3.4g. **Theorem.** *Maintaining assumptions (A1) and (A2) from §3.2,* $\dot{Z}^\nu(\sigma)$ *is a projective* $C_{\sigma,k}$*-module, for all* $\nu \vDash k$. *Moreover for any* $h \geq k$, $\bigoplus_{\nu \in \Lambda(h,k)} \dot{Z}^\nu(\sigma)$ *is a projective generator for* $\mathrm{mod}(C_{\sigma,k})$.

Proof. Fix some $h \geq k$ and set

$$Z = \bigoplus_{\nu \in \Lambda(h,k)} R^{G_n}_{G_{d\nu}} Z^\nu(\sigma) \cong \bigoplus_{\nu \in \Lambda(h,k)} \dot{Z}^\nu(\sigma),$$

$$Y = \bigoplus_{\nu \in \Lambda(h,k)} R^{G_n}_{G_{d\nu}} Y^\nu(\sigma),$$

$$Q = \bigoplus_{\nu \in \Lambda(h,k)} R^{G_n}_{G_{d\nu}} \Gamma_{d\nu}.$$

As $Y^\nu(\sigma)$ is a non-zero submodule of $Z^\nu(\sigma)$, it contains the simple socle $M^\nu(\sigma)y_\nu$ of $Z^\nu(\sigma)$ as a submodule (see Lemma 3.3b). Applying $R^{G_n}_{G_{d\nu}}$ to the inclusions $M^\nu(\sigma)y_\nu \subseteq Y^\nu(\sigma) \subseteq Z^\nu(\sigma)$, we deduce that

$$M^k(\sigma)y_\nu \subseteq R^{G_n}_{G_{d\nu}} Y^\nu(\sigma) \subseteq R^{G_n}_{G_{d\nu}} Z^\nu(\sigma)$$

as naturally embedded submodules of $M^k(\sigma)$. So, $\bigoplus_{\nu\in\Lambda(h,k)} M^k(\sigma)y_\nu \subseteq Y \subseteq Z$. Also observe that Y is a quotient of Q, since each $Y^\nu(\sigma)$ is a quotient of $\Gamma_{d\nu}$.

It follows easily from Lemma 3.4f (on expanding the direct sums) that all of

$$\mathrm{Hom}_{FG_n}(Y,Y), \mathrm{Hom}_{FG_n}(Q,Y), \mathrm{Hom}_{FG_n}(Q,Z)$$

have the same dimension, namely, $\sum_{\lambda,\mu\in\Lambda(h,k)} |D_{\mu,\lambda}|$ which is precisely the dimension of $S_{h,k}$ by (1.2a). Since Q is projective it contains the projective cover P of Y as a summand. Now the equality $\dim \mathrm{Hom}_{FG_n}(Y,Y) = \dim \mathrm{Hom}_{FG_n}(Q,Y)$ implies that $\dim \mathrm{Hom}_{FG_n}(Y,Y) = \dim \mathrm{Hom}_{FG_n}(P,Y)$. This verifies the condition in Lemma 3.2a, showing that Y is a projective $FG_n/\operatorname{ann}_{FG_n}(Y)$-module.

Now we compute $\mathrm{End}_{FG_n}(Y)$. The fact that $\mathrm{Hom}_{FG_n}(Q,Y)$ and $\mathrm{Hom}_{FG_n}(Q,Z)$ have the same dimension implies that every FG_n-homomorphism from Q to Z has image lying in Y. So since Y is certainly a quotient of Q, we can describe Y alternatively as the subspace of Z spanned by the images of all FG_n-homomorphisms from Q to Z. This alternative description makes it clear that Y is stable under all FG_n-endomorphisms of Z. So, restriction gives a well-defined map

$$\mathrm{End}_{FG_n}(Z) \to \mathrm{End}_{FG_n}(Y).$$

It is injective since we know from Theorem 3.4c and (3.4b) that the homomorphism $\mathrm{End}_{FG_n}(Z) \to \mathrm{End}_{FG_n}(\bigoplus_{\nu\in\Lambda(h,k)} M^k(\sigma)y_\nu)$ induced by restriction is injective. Since $\mathrm{End}_{FG_n}(Z) \cong S_{h,k}$ and $\mathrm{End}_{FG_n}(Y)$ has the same dimension as $S_{h,k}$, we deduce that $\mathrm{End}_{FG_n}(Y) \cong S_{h,k}$.

For $h \geq k$, the algebra $S_{h,k}$ has Par_k non-isomorphic irreducible modules. Combining the previous two paragraphs and Fitting's lemma, we deduce that Y has precisely Par_k non-isomorphic irreducible modules appearing in its head. Since Y is a direct sum of submodules of $M^k(\sigma)$, assumption (A2) now gives that every irreducible constituent of $M^k(\sigma)$ appears in the head of Y. Hence every irreducible constituent of $M^k(\sigma)$ appears in the head of the projective FG_n-module Q. Now we know that every homomorphism from Q to Z has image lying in Y, while every composition factor of Z/Y appears in the head of the projective module Q. This shows $Z = Y$.

Then, observe that $M^k(\sigma)$ is a summand of $Y = Z$, so

$$\operatorname{ann}_{FG_n}(Y) = \operatorname{ann}_{FG_n}(M^k(\sigma)).$$

In other words, $FG_n/\operatorname{ann}_{FG_n}(Y) = C_{\sigma,k}$. We have already shown that Y is a projective $FG_n/\operatorname{ann}_{FG_n}(Y)$-module, which means that Z and all its summands are projective $C_{\sigma,k}$-modules. Taking h large enough, this shows in particular that $R^{G_n}_{G_{d\nu}} Z^\nu(\sigma) \cong \dot{Z}^\nu(\sigma)$ is projective for each $\nu \vDash k$.

It remains to show that every irreducible $C_{\sigma,k}$-module appears in the head of Z. As $M^k(\sigma)$ is a faithful $C_{\sigma,k}$-module, every irreducible $C_{\sigma,k}$-module appears as some composition factor of $M^k(\sigma)$. Now we know that a copy of every composition factor of $M^k(\sigma)$ does appear in the head of Z. □

We also record at this point the following fact obtained in the proof of the theorem:

(3.4h) *The submodules $Z^k(\sigma)$ and $Y^k(\sigma)$ of $M^k(\sigma)$ coincide. So $Z^k(\sigma)$ can be characterized as the image of any non-zero homomorphism from Γ_n to $M^k(\sigma)$.*

In view of (3.4h), we will not need the notation $Y^k(\sigma)$ again.

3.5. Standard modules

Choose an integer $h \geq k$ and let $S_{h,k} = S_{F,q^d}(h,k)$. Let $Z = \bigoplus_{\nu \in \Lambda(h,k)} \dot{Z}^\nu(\sigma)$. We always regard Z as a $(C_{\sigma,k}, S_{h,k})$-bimodule, with $S_{h,k}$ acting as in Theorem 3.4c. Define the functors

$$\alpha_{\sigma,h,k} : \mathrm{mod}(C_{\sigma,k}) \to \mathrm{mod}(S_{h,k}), \qquad \alpha_{\sigma,h,k} = \mathrm{Hom}_{C_{\sigma,k}}(Z, ?), \tag{3.5.1}$$

$$\beta_{\sigma,h,k} : \mathrm{mod}(S_{h,k}) \to \mathrm{mod}(C_{\sigma,k}), \qquad \beta_{\sigma,h,k} = Z \otimes_{S_{h,k}} ?. \tag{3.5.2}$$

Because of Theorem 3.4c, Theorem 3.4g and our standing assumptions (A1) and (A2) on σ, Z is a projective generator for $\mathrm{mod}(C_{\sigma,k})$ with endomorphism algebra $S_{h,k}$, so:

(3.5a) *The functors $\alpha_{\sigma,h,k}$ and $\beta_{\sigma,h,k}$ are mutually inverse equivalences of categories.*

Recall now the basic facts about the representation theory of $S_{h,k}$ described in §1.2. In particular, $S_{h,k}$ is a quasi-hereditary algebra with weight poset $\Lambda^+(h,k)$ partially ordered by $\leq$. Also, for $\lambda \in \Lambda^+(h,k)$, we have the $S_{h,k}$-modules $L_h(\lambda), \Delta_h(\lambda)$ and $\nabla_h(\lambda)$. For $\lambda \vdash k$, we can regard its transpose λ' as an element of $\Lambda^+(h,k)$, since $h \geq k$. Define the $C_{\sigma,k}$-modules (hence FG_n-modules, inflating in the usual way):

$$L(\sigma, \lambda) = \beta_{\sigma,h,k}(L_h(\lambda')), \tag{3.5.3}$$

$$\Delta(\sigma, \lambda) = \beta_{\sigma,h,k}(\Delta_h(\lambda')), \tag{3.5.4}$$

$$\nabla(\sigma, \lambda) = \beta_{\sigma,h,k}(\nabla_h(\lambda')), \tag{3.5.5}$$

for any partition $\lambda \vdash k$. For example, as we shall see shortly, if $\sigma = 1$ then $L(1,(n)) = \Delta(1,(n)) = \nabla(1,(n))$ is the trivial FG_n-module, while the (not necessarily irreducible) modules $\Delta(1,(1^n))$ and $\nabla(1,(1^n))$ are modular reductions of the Steinberg module.

Since $\beta_{\sigma,h,k}$ is a Morita equivalence, we see at once that the algebra $C_{\sigma,k}$ is a quasi-hereditary algebra with weight poset $\{\lambda \vdash k\}$ partially ordered by $\geq$ (the opposite order to $S_{h,k}$ since we have transposed partitions). Moreover, $\{L(\sigma,\lambda)\}$, $\{\Delta(\sigma,\lambda)\}$ and $\{\nabla(\sigma,\lambda)\}$ for all $\lambda \vdash k$ give the irreducible, standard and costandard $C_{\sigma,k}$-modules. Recall also that FG_n possesses the anti-automorphism τ, induced by taking transpose matrices. Since $M^k(\sigma)^\tau \cong M^k(\sigma)$ by Corollary 2.2f, τ factors to induce an anti-automorphism of the quotient $C_{\sigma,k} = FG_n / \mathrm{ann}_{FG_n}(M^k(\sigma))$. So we also have a notion of contravariant duality on $\mathrm{mod}(C_{\sigma,k})$, and moreover, as this is true even as FG_n-modules, $L(\sigma,\lambda)^\tau \cong L(\sigma,\lambda)$ for each $\lambda \vdash k$. The following basic facts now follow immediately from the Morita equivalence or from standard properties of quasi-hereditary algebras (cf. (1.2d)):

(3.5b) (i) $\Delta(\sigma,\lambda)$ *has simple head isomorphic to* $L(\sigma,\lambda)$, *and all other composition factors are of the form* $L(\sigma,\mu)$ *for* $\mu > \lambda$.
(ii) *For* $\lambda, \mu \vdash k$, $[\Delta(\sigma,\lambda) : L(\sigma,\mu)] = [\Delta_h(\lambda') : L_h(\mu')]$.
(iii) *For* $\lambda \vdash k$, $L(\sigma,\lambda)^\tau \cong L(\sigma,\lambda)$ *and* $\Delta(\sigma,\lambda)^\tau \cong \nabla(\sigma,\lambda)$.

We pause to explain why the definitions (3.5.3)–(3.5.5) are independent of the particular choice of $h \geq k$. Take $h \geq l \geq k$ and, as explained in §1.5, identify $S_{l,k}$ with the subring $eS_{h,k}e$ of $S_{h,k}$, where $e = e_{h,l}$ is the idempotent of (1.5.1). Recall the equivalence of categories $\mathrm{infl}_{S_{l,k}}^{S_{h,k}} : \mathrm{mod}(S_{l,k}) \to \mathrm{mod}(S_{h,k})$ from (1.5.3) and (1.5a).

3.5c. **Lemma.** *The functors*

$$\beta_{\sigma,h,k} \circ \mathrm{infl}_{S_{l,k}}^{S_{h,k}} : \mathrm{mod}(S_{l,k}) \to \mathrm{mod}(C_{\sigma,k}) \quad \textit{and} \quad \beta_{\sigma,l,k} : \mathrm{mod}(S_{l,k}) \to \mathrm{mod}(C_{\sigma,k})$$

are isomorphic.

Proof. The module $\bigoplus_{\lambda\in\Lambda(l,k)} \dot{Z}^\lambda(\sigma)$ is precisely the $(C_{\sigma,k}, S_{l,k})$-subbimodule Ze of Z. So, the functor $\beta_{\sigma,l,k}$ is by definition the functor $Ze\otimes_{eS_{h,k}e}?$. Now associativity of tensor product gives the isomorphism

$$Z \otimes_{S_{h,k}} (S_{h,k}e \otimes_{eS_{h,k}e} M) \cong (Z \otimes_{S_{h,k}} S_{h,k}e) \otimes_{eS_{h,k}e} M \cong Ze \otimes_{eS_{h,k}e} M$$

for any $M \in \mathrm{mod}(S_{l,k})$. The isomorphism is clearly functorial. □

Now by (1.5a) and (1.5b), $L_h(\lambda') \cong \mathrm{infl}_{S_{l,k}}^{S_{h,k}} L_l(\lambda')$. So Lemma 3.5c shows immediately that $\beta_{\sigma,l,k}(L_l(\lambda')) \cong \beta_{\sigma,h,k}(L_h(\lambda'))$ as $C_{\sigma,k}$-modules. Hence, the definition (3.5.3) is independent of the choice of h, and the same argument gives independence of h for (3.5.4) and (3.5.5).

The next goal is to give two alternative definitions of the standard module $\Delta(\sigma,\lambda)$ without reference to the Schur algebra. Since $L(\sigma,\lambda)$ is the simple head of $\Delta(\sigma,\lambda)$, this also gives a more implicit realization of irreducibles. First, recalling the definitions (1.3.3) and (1.3.4), we have the following lemma which motivates our choice of notation:

3.5d. **Lemma.** *For* $\nu \vDash k$, $\dot{Z}^\nu(\sigma) \cong \beta_{\sigma,h,k}(Z^\nu(V_h))$ *and* $\dot{\Lambda}^\nu(\sigma) \cong \beta_{\sigma,h,k}(\Lambda^\nu(V_h))$.

Proof. By (1.3b)(i), $Z^\nu(V_h)$ is isomorphic to the left ideal $S_{h,k}\phi^1_{\nu,\nu}$ of $S_{h,k}$. By Lemma 3.1f, $\beta_{\sigma,h,k}(S_{h,k}\phi^1_{\nu,\nu}) \cong Z\phi^1_{\nu,\nu}$, which is precisely the summand $\dot{Z}^\nu(\sigma)$ of Z by the definition of the action of $\phi^1_{\nu,\nu}$ from Theorem 3.4c.

Similarly, by (1.3b)(ii), $\Lambda^\nu(V_h) \cong S_{h,k}\kappa(y_\nu)$, so $\beta_{\sigma,h,k}(S_{h,k}\kappa(y_\nu)) \cong Z\kappa(y_\nu)$. By definition of the embedding κ from (1.2b), together with (1.1c) and Theorem 3.4c,

$$Z\kappa(y_\nu) = M^k(\sigma)y_\nu^\# = M^k(\sigma)x_\nu = \dot{\Lambda}^\nu(\sigma)$$

as required. □

Now we obtain the desired characterizations of $\Delta(\sigma,\lambda)$. Recall the definition of $u_\lambda \in \Sigma_k$ from (1.1d).

3.5e. **Theorem.** *For $\lambda \vdash k$,*

(i) *the space* $\mathrm{Hom}_{C_{\sigma,k}}(\dot{Z}^{\lambda'}(\sigma), \dot{\Lambda}^\lambda(\sigma))$ *is one dimensional, and the image of any non-zero such homomorphism is isomorphic to $\Delta(\sigma,\lambda)$;*

(ii) $\Delta(\sigma,\lambda)$ *is isomorphic to the submodule $\dot{Z}^{\lambda'}(\sigma)T_{u_\lambda}x_\lambda$ of $M^k(\sigma)$.*

Proof. (i) This is immediate from Lemma 3.5d, the definition (3.5.4) and (1.3d), since $\beta_{\sigma,h,k}$ is an equivalence of categories.

(ii) First observe that $\dot{Z}^{\lambda'}(\sigma)T_{u_\lambda}x_\lambda$ is both a homomorphic image of $\dot{Z}^{\lambda'}(\sigma)$ and a submodule of $\Lambda^\lambda(\sigma)$. So in view of (i), the result will follow once we show that $\dot{Z}^{\lambda'}(\sigma)T_{u_\lambda}x_\lambda$ is non-zero. Well, $\dot{Z}^{\lambda'}(\sigma)$ contains $M^k(\sigma)y_{\lambda'}$ as a submodule. Moreover, $M^k(\sigma)y_{\lambda'}T_{u_\lambda}x_\lambda$ is non-zero, as $M^k(\sigma)$ is a faithful H_k-module and $y_{\lambda'}T_{u_\lambda}x_\lambda \neq 0$ by (1.1d). □

3.5f. **Remark.** In [J2, Definition 7.7], James defines right FG_n-modules $S(\sigma,\lambda)$ for each $\lambda \vdash k$. In view of (3.4h), Theorem 3.5e(ii) is a left module analogue of James' definition.

As we did in §1.2, we will write $\tilde{M}$ for the right $C_{\sigma,k}$-module obtained from $M \in \mathrm{mod}(C_{\sigma,k})$ by twisting the left action into a right action using τ. In this way, we obtain right $C_{\sigma,k}$-modules $\tilde{\Delta}(\sigma,\lambda), \tilde{L}(\sigma,\lambda)$ and $\tilde{\nabla}(\sigma,\lambda)$. Now we conclude the section with some extensions of (1.2e):

3.5g. **Theorem.** (i) *$C_{\sigma,k}$ has a filtration as a $(C_{\sigma,k}, C_{\sigma,k})$ bimodule with factors isomorphic to $\Delta(\sigma,\lambda) \otimes \tilde{\Delta}(\sigma,\lambda)$, each appearing precisely once for each $\lambda \vdash k$ and ordered in any way refining the dominance order on partitions so that factors corresponding to most dominant λ appear at the top of the filtration.*

(ii) $Z = \bigoplus_{\nu\in\Lambda(h,k)} \dot{Z}^\nu(\sigma)$ *has a filtration as a $(C_{\sigma,k}, S_{h,k})$-bimodule with factors $\Delta(\sigma,\lambda) \otimes \tilde{\Delta}_h(\lambda')$ appearing precisely once for each $\lambda \vdash k$ and ordered in any way refining the dominance order so that factors corresponding to most dominant λ appear at the top of the filtration.*

Proof. (i) This follows immediately from the general theory of quasi-hereditary algebras, in the same way as explained after (1.2e).

(ii) The functor $Z \otimes_{S_{h,k}} ?$ can also be viewed as an exact functor from the category of $(S_{h,k}, S_{h,k})$-bimodules to the category of $(C_{\sigma,k}, S_{h,k})$-bimodules. Clearly,

$$\begin{aligned} Z \otimes_{S_{h,k}} (\Delta_h(\lambda) \otimes \tilde{\Delta}_h(\lambda)) &\cong (Z \otimes_{S_{h,k}} \Delta_h(\lambda)) \otimes \tilde{\Delta}_h(\lambda) \\ &\cong \Delta(\sigma, \lambda') \otimes \tilde{\Delta}_h(\lambda). \end{aligned}$$

So now applying $Z \otimes_{S_{h,k}} ?$ to the filtration of (1.2e) gives the result. $\square$

Chapter 4

Further connections and applications

In this chapter, we prove a number of results that supplement the main Morita theorem of the previous chapter. In particular, we make precise the idea that tensor products in the quantum linear group correspond to Harish-Chandra induction under the Morita equivalence, and extend the Morita theorem both to the ground ring $\mathcal{O}$ and to p-singular elements σ.

4.1. Base change

Let $\sigma \in \bar{\mathbb{F}}_q^\times$ be of degree d over $\mathbb{F}_q$ satisfying the conditions (A1) and (A2) from §3.2. Write $n = kd$ for some $k \geq 1$ and let R denote one of the rings F, K or $\mathcal{O}$. Recall that we have defined the $(RG_n, H_{k,R})$-bimodule $M^k(\sigma)_R$ in (2.4.2), with $M^k(\sigma) = M^k(\sigma)_F$. Define

$$Z^k(\sigma)_R = \{m \in M^k(\sigma)_R \mid mh = \mathcal{E}_{H_{k,R}}(h)m \text{ for all } h \in H_{k,R}\},$$

so our original module $Z^k(\sigma)$ from (3.3.2) is precisely $Z^k(\sigma)_F$.

4.1a. **Lemma.** *$Z^k(\sigma)_\mathcal{O}$ is an $\mathcal{O}$-free $\mathcal{O}$-module of finite rank, with*

$$K \otimes_\mathcal{O} Z^k(\sigma)_\mathcal{O} \cong Z^k(\sigma)_K,$$
$$F \otimes_\mathcal{O} Z^k(\sigma)_\mathcal{O} \cong Z^k(\sigma)_F.$$

Proof. Recalling that $M^k(\sigma)_\mathcal{O}$ is an $\mathcal{O}$-lattice in $M^k(\sigma)_K$, we have by definition that

$$Z^k(\sigma)_\mathcal{O} = Z^k(\sigma)_K \cap M^k(\sigma)_\mathcal{O}.$$

It is immediate from this that $Z^k(\sigma)_\mathcal{O}$ is an $\mathcal{O}$-lattice in $Z^k(\sigma)_K$, and also that it is a pure submodule of $M^k(\sigma)_\mathcal{O}$ (see [La, 17.1(i)]). Consequently, the natural map $i : F \otimes_\mathcal{O} Z^k(\sigma)_\mathcal{O} \to M^k(\sigma)_F$ induced by the embedding $Z^k(\sigma)_\mathcal{O} \hookrightarrow M^k(\sigma)_\mathcal{O}$ is also an embedding. Clearly, since the action of H_k on $M^k(\sigma)$ is compatible with base change, the image of i is contained in $Z^k(\sigma)_F$.

Now for $R = F, K$ or $\mathcal{O}$, let $Y^k(\sigma)_R$ denote the submodule of $M^k(\sigma)_R$ spanned by the images of all RG_n-homomorphisms from $\Gamma_{n,R}$ to $Z^k(\sigma)_R$. Since Γ_n is projective, any FG_n-homomorphism from Γ_n to $M^k(\sigma)$ is induced by base change from some $\mathcal{O}G_n$-homomorphism from $\Gamma_{n,\mathcal{O}}$ to $M^k(\sigma)_{\mathcal{O}}$. It follows directly that the natural (not necessarily injective) map from $F \otimes_{\mathcal{O}} Y^k(\sigma)_{\mathcal{O}}$ to $M^k(\sigma)_F$ has image $Y^k(\sigma)_F$. Now by Corollary 2.5e(iii), $Y^k(\sigma)_{\mathcal{O}} \subseteq Z^k(\sigma)_{\mathcal{O}}$, so we deduce that the image of i contains $Y^k(\sigma)_F$.

Finally, we appeal to (3.4h), where we observed that $Y^k(\sigma)_F = Z^k(\sigma)_F$. So the two previous paragraphs show in fact that $i : F \otimes_{\mathcal{O}} Z^k(\sigma)_{\mathcal{O}} \to Z^k(\sigma)_F$ is an isomorphism. □

For $\nu \vDash k$, let $H_{\nu,R}$ denote the parabolic subalgebra of $H_{k,R}$ with sign representation $\mathcal{E}_{H_{\nu,R}}$. Define

$$\dot{Z}^\nu(\sigma)_R = \{m \in M^k(\sigma)_R \mid mh = \mathcal{E}_{H_{\nu,R}}(h)m \text{ for all } h \in H_{\nu,R}\}, \tag{4.1.1}$$

so our original module $\dot{Z}^\nu(\sigma)$ from (3.3.5) is precisely $\dot{Z}^\nu(\sigma)_F$. For $\nu = (k_1, \dots, k_a)$, $\dot{Z}^\nu(\sigma) \cong R^{G_n}_{G_{d\nu}} Z_{k_1}(\sigma) \boxtimes \cdots \boxtimes Z_{k_a}(\sigma)$ (cf. the comments after (3.3.6)). So since Harish-Chandra induction commutes with base change, we deduce from Lemma 4.1a that:

(4.1b) *$\dot{Z}^\nu(\sigma)_{\mathcal{O}}$ is an $\mathcal{O}$-free $\mathcal{O}$-module of finite rank, with*

$$K \otimes_{\mathcal{O}} \dot{Z}^\nu(\sigma)_{\mathcal{O}} \cong \dot{Z}^\nu(\sigma)_K,$$
$$F \otimes_{\mathcal{O}} \dot{Z}^\nu(\sigma)_{\mathcal{O}} \cong \dot{Z}^\nu(\sigma)_F.$$

Now we can define standard modules over $\mathcal{O}$. For $R = F, K$ or $\mathcal{O}$ and $\lambda \vDash k$, set

$$\Delta(\sigma, \lambda)_R = \dot{Z}^{\lambda'}(\sigma)_R T_{u_\lambda} x_\lambda.$$

By Theorem 3.5e, $\Delta(\sigma, \lambda)_F \cong \Delta(\sigma, \lambda)$.

4.1c. **Theorem.** *$\Delta(\sigma, \lambda)_{\mathcal{O}}$ is $\mathcal{O}$-free of finite rank with*

$$K \otimes_{\mathcal{O}} \Delta(\sigma, \lambda)_{\mathcal{O}} \cong \Delta(\sigma, \lambda)_K,$$
$$F \otimes_{\mathcal{O}} \Delta(\sigma, \lambda)_{\mathcal{O}} \cong \Delta(\sigma, \lambda)_F.$$

Moreover, the character of the KG_n-module $\Delta(\sigma, \lambda)_K$ is precisely $\chi_{\sigma,\lambda}$, so $\Delta(\sigma, \lambda)$ is the reduction modulo p of a KG_n-module affording the character $\chi_{\sigma,\lambda}$.

Proof. First observe that $\Delta(\sigma, \lambda)_{\mathcal{O}}$ is torsion free as it is a submodule of the torsion free $\mathcal{O}$-module $M^k(\sigma)_{\mathcal{O}}$. There is a natural map $K \otimes_{\mathcal{O}} \Delta(\sigma, \lambda)_{\mathcal{O}} \to M^k(\sigma)_K$, which is injective as K is flat over $\mathcal{O}$. One easily checks that its image is precisely

$\Delta(\sigma,\lambda)_K$, hence proving that $\Delta(\sigma,\lambda)_{\mathcal{O}}$ is an $\mathcal{O}$-lattice in $\Delta(\sigma,\lambda)_K$. Hence, in particular, $\Delta(\sigma,\lambda)_{\mathcal{O}}$ has rank equal to $\dim \Delta_k(\sigma,\lambda)_K$. Now applying Theorem 3.5g(ii) (or rather, its easier analogue over K), taking $h = k$ for definiteness, we see that

$$\sum_{\nu\in\Lambda(k,k)} \dim \dot{Z}^\nu(\sigma)_K = \sum_{\lambda\vdash k} (\dim \Delta(\sigma,\lambda)_K)(\dim \Delta_k(\lambda')_K)$$

where $\Delta_k(\lambda')_K$ denotes the (irreducible) standard module for the Schur algebra algebra $S_{K,q^d}(k,k)$ over K. Using (4.1b), $\dim \dot{Z}^\nu(\sigma)_K = \dim \dot{Z}^\nu(\sigma)_F$, while it is well-known that the dimension of standard modules for the Schur algebra do not depend on the ground field. So, we see that

$$\sum_{\nu\in\Lambda(k,k)} \dim \dot{Z}^\nu(\sigma)_F = \sum_{\lambda\vdash k} (\dim \Delta(\sigma,\lambda)_K)(\dim \Delta_k(\lambda')_F).$$

Now again, there is a natural map $i : F \otimes_{\mathcal{O}} \Delta(\sigma,\lambda)_{\mathcal{O}} \to M^k(\sigma)_{\mathcal{O}}$ induced by the embedding $\Delta(\sigma,\lambda)_{\mathcal{O}} \hookrightarrow M^k(\sigma)_{\mathcal{O}}$, with image $\Delta(\sigma,\lambda)_F$. This shows that $\dim \Delta(\sigma,\lambda)_K \geq \dim \Delta(\sigma,\lambda)_F$. On the other hand, applying Theorem 3.5g(ii) over F, we have that

$$\sum_{\nu\in\Lambda(k,k)} \dim \dot{Z}^\nu(\sigma)_F = \sum_{\lambda\vdash k} (\dim \Delta(\sigma,\lambda)_F)(\dim \Delta_k(\lambda')_F).$$

Comparing with our previous expression, we see that $\dim \Delta(\sigma,\lambda)_K$ must actually equal $\dim \Delta(\sigma,\lambda)_F$ for all $\lambda \vdash k$, hence that i is injective.

It remains to explain why $\Delta(\sigma,\lambda)_K$ has character $\chi_{\sigma,\lambda}$. Write $\lambda = (l_1,\dots,l_a)$ and $\lambda' = (l'_1,\dots,l'_b)$. First observe by (1.1e) and (2.3f) that $\chi_{\sigma,\lambda}$ is the unique irreducible character that is a constituent of both of the characters $R^{G_n}_{G_{d\lambda}}(\chi_{\sigma,(l_1)}\cdots\chi_{\sigma,(l_a)})$ and $R^{G_n}_{G_{d\lambda'}}(\chi_{\sigma,(1^{l'_1})}\cdots\chi_{\sigma,(1^{l'_a})})$. So recalling Lemma 2.5c, the modules

$$M^k(\sigma)_K x_\lambda \cong R^{G_n}_{G_{d\lambda}}(M^\lambda(\sigma)_K x_\lambda),$$
$$M^k(\sigma)_K y_{\lambda'} \cong R^{G_n}_{G_{d\lambda'}}(M^{\lambda'}(\sigma)_K y_{\lambda'})$$

have a unique irreducible composition factor in common, with character $\chi_{\sigma,\lambda}$. But in characteristic zero, $\dot{Z}^{\lambda'}(\sigma)_K = M^k(\sigma)_K y_{\lambda'}$ as then $y_{\lambda'}$ is an idempotent (up to a non-zero scalar). So by definition, $\Delta(\sigma,\lambda)_K = M^k(\sigma)_K y_{\lambda'} T_{u_\lambda} x_\lambda$ so is an irreducible quotient of $M^k(\sigma)_K y_{\lambda'}$ and an irreducible submodule of $M^k(\sigma)_K x_\lambda$. So $\Delta(\sigma,\lambda)_K$ does indeed have character $\chi_{\sigma,\lambda}$. □

Using Theorem 4.1c, we can extend all our earlier results to the ground ring $\mathcal{O}$. For $R = F, K$ or $\mathcal{O}$, set

$$C_{\sigma,k,R} = C_{R,(\sigma)^k}(GL_n(\mathbb{F}_q)) = RG_n / \operatorname{ann}_{RG_n}(M^k(\sigma)_R).$$

So, $C_{\sigma,k,F}$ is precisely the algebra $C_{\sigma,k}$ as defined in (3.2.1). We call $C_{\sigma,k,R}$ the *cuspidal algebra over* R. Note that if σ is p-regular, $C_{\sigma,k,R}$ is in fact a quotient of the block algebra $B_{\sigma,k,R}$ introduced at the end of §2.4.

4.1d. **Theorem.** *Maintaining the assumptions (A1) and (A2) on σ from §3.2,*

(i) *$C_{\sigma,k,\mathcal{O}}$ is $\mathcal{O}$-free of finite rank with $K \otimes_{\mathcal{O}} C_{\sigma,k,\mathcal{O}} \cong C_{\sigma,k,K}$ and $F \otimes_{\mathcal{O}} C_{\sigma,k,\mathcal{O}} \cong C_{\sigma,k,F}$;*

(ii) *for each $\nu \models k$, $\dot{Z}^\nu(\sigma)_{\mathcal{O}}$ is a projective $C_{\sigma,k,\mathcal{O}}$-module;*

(iii) *the endomorphism algebra $\mathrm{End}_{C_{\sigma,k,\mathcal{O}}}(\bigoplus_{\nu\in\Lambda(h,k)} \dot{Z}^\nu(\sigma)_{\mathcal{O}})$ is isomorphic to the q^d-Schur algebra $S_{\mathcal{O},q^d}(h,k)$;*

(iv) *for $h \geq k$, $\bigoplus_{\nu\in\Lambda(h,k)} \dot{Z}^\nu(\sigma)_{\mathcal{O}}$ is a projective generator for $\mathrm{mod}(C_{\sigma,k,\mathcal{O}})$, so $C_{\sigma,k,\mathcal{O}}$ is Morita equivalent to $S_{\mathcal{O},q^d}(h,k)$.*

Proof. (i) By definition, $C_{\sigma,k,R}$ is the R-submodule of $\mathrm{End}_R(M^k(\sigma)_R)$ spanned by the images of the elements of G_n. So it is finitely generated, $C_{\sigma,k,\mathcal{O}}$ is contained in $C_{\sigma,k,K}$ and spans $C_{\sigma,k,K}$ over K. This shows that $C_{\sigma,k,\mathcal{O}}$ is an $\mathcal{O}$-lattice in $C_{\sigma,k,K}$. Tensoring the inclusion $C_{\sigma,k,\mathcal{O}} \hookrightarrow \mathrm{End}_{\mathcal{O}}(M^k(\sigma)_{\mathcal{O}})$ with F, we obtain a natural map $F \otimes_{\mathcal{O}} C_{\sigma,k,\mathcal{O}} \to F \otimes_{\mathcal{O}} \mathrm{End}_{\mathcal{O}}(M^k(\sigma)_{\mathcal{O}}) \cong \mathrm{End}_F(M^k(\sigma)_F)$ whose image is clearly $C_{\sigma,k,F}$. To show that this surjection is injective, we need to check that $\dim C_{\sigma,k,K} = \dim C_{\sigma,k,F}$. But by Theorem 3.5g(i), its analogue over K and Theorem 4.1c:

$$\dim C_{\sigma,k,F} = \sum_{\lambda\vdash k} (\dim \Delta(\sigma,\lambda)_F)^2 = \sum_{\lambda\vdash k} (\dim \Delta(\sigma,\lambda)_K)^2 = \dim C_{\sigma,k,K}.$$

(ii) As $\dot{Z}^\nu(\sigma)_F$ is a projective $C_{\sigma,k,F}$-module from Theorem 3.4g, we see from (i) and lifting idempotents (e.g. see [La, 14.4]) that it has a unique lift to a projective $C_{\sigma,k,\mathcal{O}}$-module. This must be $\dot{Z}^\nu(\sigma)_{\mathcal{O}}$ thanks to (4.1b).

(iii) Write E_R for $\mathrm{End}_{C_{\sigma,k,R}}\left(\bigoplus_{\nu\in\Lambda(h,k)} \dot{Z}^\nu(\sigma)_R\right)$. We know by Theorem 3.4c (or its analogue over K) that $E_F \cong S_{F,q^d}(h,k)$ and $E_K \cong S_{K,q^d}(h,k)$. Moreover, $E_{\mathcal{O}}$ is an $\mathcal{O}$-lattice in E_K and there is a natural embedding $F \otimes_{\mathcal{O}} E_{\mathcal{O}} \hookrightarrow E_F$ which is an isomorphism by dimension. So we can identify $K \otimes_{\mathcal{O}} E_{\mathcal{O}}$ with E_K and $F \otimes_{\mathcal{O}} E_{\mathcal{O}}$ with E_F

Now, the basis element $\phi^u_{\mu,\lambda}$ of $E_K \cong S_{K,q^d}(h,k)$ acts as zero on all summands except $\dot{Z}^\mu(\sigma)_K$ where it is induced by right multiplication by $h = \sum_{w\in\Sigma_\mu u\Sigma_\lambda\cap D_\mu^{-1}} T_w^{\#}$. By (4.1.1), $\dot{Z}^\nu(\sigma)_{\mathcal{O}} = \dot{Z}^\nu(\sigma)_K \cap M^k(\sigma)_{\mathcal{O}}$. Also, h lies in $H_{\mathcal{O},q^d}(\Sigma_k)$ so stabilizes $M^k(\sigma)_{\mathcal{O}}$. Hence, $\dot{Z}^\mu(\sigma)_{\mathcal{O}} h \subseteq \dot{Z}^\lambda(\sigma)_{\mathcal{O}}$, so each $\phi^u_{\mu,\lambda} \in E_K$ restricts to give a well-defined element of $E_{\mathcal{O}}$. We have constructed an isomorphic copy $S_{\mathcal{O}}$ of $S_{\mathcal{O},q^d}(h,k)$ in $E_{\mathcal{O}}$, namely, the $\mathcal{O}$-span of the standard basis elements $\phi^u_{\mu,\lambda} \in S_{K,q^d}(h,k)$.

It remains to show that $S_{\mathcal{O}} = E_{\mathcal{O}}$. We have a short exact sequence $0 \to S_{\mathcal{O}} \to E_{\mathcal{O}} \to Q_{\mathcal{O}} \to 0$ of $\mathcal{O}$-modules. To prove that $Q_{\mathcal{O}} = 0$, it suffices to show that $F \otimes_{\mathcal{O}} Q_{\mathcal{O}} = 0$. Tensoring with F, we have an exact sequence

$$F \otimes_{\mathcal{O}} S_{\mathcal{O}} \xrightarrow{i} E_F \longrightarrow F \otimes_{\mathcal{O}} Q_{\mathcal{O}} \longrightarrow 0.$$

Now, the map i sends $1 \otimes \phi^u_{\mu,\lambda}$ to the corresponding endomorphism $\phi^u_{\mu,\lambda}$ defined as in Theorem 3.4c. Hence, i is injective so an isomorphism by dimension. We deduce that $F \otimes_{\mathcal{O}} Q_{\mathcal{O}} = 0$ to complete the proof.

(iv) We have seen in (ii) that $\bigoplus_{\nu\in\Lambda(h,k)} \dot{Z}^\nu(\sigma)_\mathcal{O}$ is a projective $C_{\sigma,k,\mathcal{O}}$-module. It is a generator because this is so on tensoring with F, using (i) and Theorem 3.4g. The statement about Morita equivalence follows directly from this and (iii). □

4.1e. **Remark.** We remark that the fundamental theorem of Cline, Parshall and Scott [CPS3, §9] follows easily at this point from Theorem 4.1d and the results of §2.4. We refer the reader to *loc. cit.* for the precise statement, as well as for applications to cohomology of FG_n.

4.2. Connecting Harish-Chandra induction with tensor products

Let $\sigma \in \bar{\mathbb{F}}_q^\times$ be of degree d over $\mathbb{F}_q$ satisfying the conditions (A1) and (A2) from §3.2. For *any* $h \geq k \geq 1$, $S_{h,k}$ denotes the algebra $S_{F,q^d}(h,k)$ and $\beta_{\sigma,h,k}$ is the functor of (3.5.2), but regarded now as a functor from $\mathrm{mod}(S_{h,k})$ to $\mathrm{mod}(FG_{kd})$ via the evident full embedding $\mathrm{mod}(C_{\sigma,h,k}) \hookrightarrow \mathrm{mod}(FG_{kd})$. Now *fix* integers $h \geq k \geq 1$, a composition $\nu = (k_1, \dots, k_a) \vDash k$ and set $n = kd$.

The main result of the section relates tensor products in the quantum linear group to Harish-Chandra induction in the finite linear group, as follows:

4.2a. **Theorem.** *The following functors are isomorphic:*

$$R^{G_n}_{G_{d\nu}}(\beta_{\sigma,h,k_1}\,?\boxtimes\cdots\boxtimes\beta_{\sigma,h,k_a}\,?) : \mathrm{mod}(S_{h,k_1})\times\cdots\times\mathrm{mod}(S_{h,k_a}) \to \mathrm{mod}(FG_n),$$
$$\beta_{\sigma,h,k}(\,?\otimes\cdots\otimes ?\,) : \mathrm{mod}(S_{h,k_1})\times\cdots\times\mathrm{mod}(S_{h,k_a}) \to \mathrm{mod}(FG_n).$$

Proof. Choose $\mu = (h_1, \dots, h_a) \vDash h$ with each $h_i \geq k_i$. Let $S_{\mu,\nu}$ denote the algebra $S_{h_1,k_1}\otimes\cdots\otimes S_{h_a,k_a}$ for short. Write $\Lambda(\mu,\nu)$ for the set of all compositions $\gamma = (g_1,\dots,g_h) \in \Lambda(h,k)$ such that, *defining* $\gamma_1 = (g_1,\dots,g_{h_1}), \gamma_2 = (g_{h_1+1},\dots,g_{h_1+h_2}),\dots,\gamma_a = (g_{h_1+\dots+h_{a-1}+1},\dots,g_h)$, we have that $\gamma_i \in \Lambda(h_i,k_i)$ for each $i = 1,\dots,a$. Consider the set of triples:

$$\Omega = \{(\gamma,\delta,u) \mid \gamma,\delta\in\Lambda(\mu,\nu), u \in D^\nu_{\gamma,\delta}\}.$$

For a triple $(\gamma,\delta,u)\in\Omega$, so that $\gamma_i,\delta_i\in\Lambda(h_i,k_i)$ for each $i = 1,\dots,a$, we have that $u = (u_1,\dots,u_a) \in \Sigma_\nu = \Sigma_{k_1}\times\cdots\times\Sigma_{k_a}$ with each $u_i \in D_{\gamma_i,\delta_i}$; so we can associate the element

$$\bar{\phi}^u_{\gamma,\delta} = \phi^{u_1}_{\gamma_1,\delta_1}\otimes\cdots\otimes\phi^{u_a}_{\gamma_a,\delta_a}$$

of $S_{\mu,\nu}$. The set $\{\bar{\phi}^u_{\gamma,\delta} \mid (\gamma,\delta,u)\in\Omega\}$ gives a natural basis for the algebra $S_{\mu,\nu}$.

Now define

$$Z = \bigoplus_{\lambda\in\Lambda(h,k)} R^{G_n}_{G_{d\lambda}} Z^\lambda(\sigma) \cong \bigoplus_{\lambda\in\Lambda(h,k)} \dot{Z}^\lambda(\sigma),$$

$$Z_\nu = \bigoplus_{\lambda\in\Lambda(\mu,\nu)} R^{G_{d\nu}}_{G_{d\lambda}} Z^\lambda(\sigma) \cong \left[\bigoplus_{\lambda_1\in\Lambda(h_1,k_1)} \dot{Z}^{\lambda_1}(\sigma)\right]\boxtimes\cdots\boxtimes\left[\bigoplus_{\lambda_a\in\Lambda(h_a,k_a)} \dot{Z}^{\lambda_a}(\sigma)\right].$$

We regard Z as an $(FG_n, S_{h,k})$-bimodule in the usual way, and also view Z_ν as an $(FG_{d\nu}, S_{\mu,\nu})$-bimodule, where the action of $S_{\mu,\nu} = S_{h_1,k_1} \otimes \cdots \otimes S_{h_a,k_a}$ on Z_ν is as described in Theorem 3.4c for each term in the above outer tensor product.

Then, $R^{G_n}_{G_{d\nu}} Z_\nu$ is an $(FG_n, S_{\mu,\nu})$-bimodule in a natural way. Moreover, by transitivity of Harish-Chandra induction,

$$R^{G_n}_{G_{d\nu}} Z_\nu \cong \bigoplus_{\lambda \in \Lambda(\mu,\nu)} \dot{Z}^\lambda(\sigma),$$

so $R^{G_n}_{G_{d\nu}} Z_\nu$ can be identified with the summand Ze_ν of the $(FG_n, S_{h,k})$-bimodule Z, where e_ν is the idempotent

$$e_\nu = \sum_{\lambda \in \Lambda(\mu,\nu)} \phi^1_{\lambda,\lambda} \in S_{h,k}.$$

Identifying $e_\nu S_{h,k} e_\nu$ with $\mathrm{End}_{FG_n}(Ze_\nu)$, we obtain an algebra embedding of $S_{\mu,\nu}$ into $e_\nu S_{h,k} e_\nu$. By definition of the actions of $S_{\mu,\nu}$ and $e_\nu S_{h,k} e_\nu$ on Ze_ν and Lemma 3.2f(i), this embedding maps the basis element $\bar{\phi}^u_{\gamma,\delta}$ of $S_{\mu,\nu}$ to $\phi^u_{\gamma,\delta} \in e_\nu S_{h,k} e_\nu$, for all $(\gamma, \delta, u) \in \Omega$. In other words:

(4.2b) *Identifying $S_{\mu,\nu}$ with a subalgebra of $e_\nu S_{h,k} e_\nu$ via the map $\bar{\phi}^u_{\gamma,\delta} \mapsto \phi^u_{\gamma,\delta}$, the $(FG_n, S_{\mu,\nu})$-bimodule $R^{G_n}_{G_{d\nu}} Z_\nu$ is isomorphic to Ze_ν, regarding the latter as an $(FG_n, S_{\mu,\nu})$-bimodule by restricting the natural action of $e_\nu S_{h,k} e_\nu$ to $S_{\mu,\nu}$.*

Now let $S_{\mu,k}$ denote the Levi subalgebra of $S_{h,k}$ as in (1.3g). Recalling the decomposition (1.3.8), the idempotent $e_\nu \in S_{h,k}$ is precisely the central idempotent of $S_{\mu,k}$ such that $e_\nu S_{\mu,k} e_\nu$ is isomorphic to $S_{\mu,\nu}$. So in fact, the embedding of $S_{\mu,\nu}$ into $e_\nu S_{h,k} e_\nu$ from (4.2b) identifies $S_{\mu,\nu}$ with the summand $e_\nu S_{\mu,k} e_\nu$ of $S_{\mu,k}$. Making this identification, define the functor

$$I : \mathrm{mod}(S_{\mu,\nu}) \to \mathrm{mod}(S_{h,k}), \qquad I = S_{h,k} e_\nu \otimes_{e_\nu S_{\mu,k} e_\nu} ?.$$

Using associativity of tensor product, the functor I can be thought of as the composite of the natural inflation functor $S_{\mu,k} e_\nu \otimes_{e_\nu S_{\mu,k} e_\nu} ? : \mathrm{mod}(S_{\mu,\nu}) \to \mathrm{mod}(S_{\mu,k})$ followed by ordinary induction $\mathrm{ind}^{S_{h,k}}_{S_{\mu,k}} : \mathrm{mod}(S_{\mu,k}) \to \mathrm{mod}(S_{h,k})$ as defined in §1.5. In view of this, the following fundamental fact follows immediately from Theorem 1.5d:

(4.2c) *The following functors are isomorphic:*

$$I(\,?\boxtimes \cdots \boxtimes ?\,) : \mathrm{mod}(S_{h_1,k_1}) \times \cdots \times \mathrm{mod}(S_{h_a,k_a}) \to \mathrm{mod}(S_{h,k}),$$

$$\mathrm{infl}^{S_{h,k_1}}_{S_{h_1,k_1}} ? \otimes \cdots \otimes \mathrm{infl}^{S_{h,k_a}}_{S_{h_a,k_a}} ? : \mathrm{mod}(S_{h_1,k_1}) \times \cdots \times \mathrm{mod}(S_{h_a,k_a}) \to \mathrm{mod}(S_{h,k}).$$

We next claim that the diagram

$$\begin{array}{ccc} \mathrm{mod}(S_{\mu,\nu}) & \xrightarrow{Z_\nu \otimes_{S_{\mu,\nu}} ?} & \mathrm{mod}(FG_{d\nu}) \\ {\scriptstyle I}\downarrow & & \downarrow{\scriptstyle R^{G_n}_{G_{d\nu}}} \\ \mathrm{mod}(S_{h,k}) & \xrightarrow[Z \otimes_{S_{h,k}} ?]{} & \mathrm{mod}(FG_n) \end{array} \tag{4.2.1}$$

commutes, i.e. that the following Schur algebra analogue of Corollary 3.2g(i) holds:

(4.2d) *The functors* $R^{G_n}_{G_{d\nu}} \circ Z_\nu \otimes_{S_{\mu,\nu}} ?$ *and* $\beta_{\sigma,h,k} \circ I : \mathrm{mod}(S_{\mu,\nu}) \to \mathrm{mod}(FG_n)$ *are isomorphic.*

To prove (4.2d), (4.2b) and associativity of tensor product gives the natural isomorphisms

$$\begin{aligned} R^{G_n}_{G_{d\nu}}(Z_\nu \otimes_{S_{\mu,\nu}} N) &\cong (R^{G_n}_{G_{d\nu}} Z_\nu) \otimes_{S_{\mu,\nu}} N \cong Ze_\nu \otimes_{S_{\mu,\nu}} N = Ze_\nu \otimes_{e_\nu S_{\mu,k} e_\nu} N \\ &\cong Z \otimes_{S_{h,k}} S_{h,k} e_\nu \otimes_{e_\nu S_{\mu,k} e_\nu} N = \beta_{\sigma,h,k} \circ I(N) \end{aligned}$$

for any $N \in \mathrm{mod}(S_{\mu,\nu})$.

Now, we obviously have the isomorphism of functors:

$$\beta_{\sigma,h_1,k_1} ? \boxtimes \cdots \boxtimes \beta_{\sigma,h_a,k_a} ? \cong Z_\nu \otimes_{S_{\mu,\nu}} (? \boxtimes \cdots \boxtimes ?)$$

as functors from $\mathrm{mod}(S_{h_1,k_1}) \times \cdots \times \mathrm{mod}(S_{h_a,k_a})$ to $\mathrm{mod}(FG_{d\nu})$. In view of this, (4.2c) and (4.2d), we deduce:

(4.2e) *There is an isomorphism*

$$R^{G_n}_{G_{d\nu}}(\beta_{\sigma,h_1,k_1} ? \boxtimes \cdots \boxtimes \beta_{\sigma,h_a,k_a} ?) \cong \beta_{\sigma,h,k}(\mathrm{infl}^{S_{h,k_1}}_{S_{h_1,k_1}} ? \otimes \cdots \otimes \mathrm{infl}^{S_{h,k_a}}_{S_{h_a,k_a}} ?)$$

as functors from $\mathrm{mod}(S_{h_1,k_1}) \times \cdots \times \mathrm{mod}(S_{h_a,k_a})$ *to* $\mathrm{mod}(FG_n)$.

Finally, by (1.5a), the functor $\mathrm{infl}^{S_{h,k_i}}_{S_{h_i,k_i}} : \mathrm{mod}(S_{h_i,k_i}) \to \mathrm{mod}(S_{h,k_i})$ is an equivalence of categories, for each $i = 1, \dots, a$. By Lemma 3.5c, the functors $\beta_{\sigma,h,k_i} \circ \mathrm{infl}^{S_{h,k_i}}_{S_{h_i,k_i}}$ and β_{σ,h_i,k_i} are isomorphic. The theorem follows on combining these statements and (4.2e). $\square$

We now give an immediate application of Theorem 4.2a. Say that a $C_{\sigma,k}$-module M has a Δ-*filtration* (resp. a ∇-*filtration*) if it has a filtration

$$0 = M_0 < M_1 < \cdots < M_b = M$$

such that for each $i = 1, \dots, b$, the factor M_i/M_{i-1} is isomorphic to $\Delta(\sigma, \lambda)$ (resp. $\nabla(\sigma, \lambda)$) for some partition $\lambda \vdash k$ (depending on i). Analogously, we say that a $C_{\sigma,\nu}$-module M has a Δ-filtration (resp. a ∇-filtration) if it has a filtration

$$0 = M_0 < M_1 < \cdots < M_b = M$$

such that for each $i = 1, \dots, b$, the factor M_i/M_{i-1} is isomorphic to $\Delta(\sigma, \lambda_1) \boxtimes \cdots \boxtimes \Delta(\sigma, \lambda_a)$ (resp. $\nabla(\sigma, \lambda_1) \boxtimes \cdots \boxtimes \nabla(\sigma, \lambda_a)$) for some partitions $\lambda_1 \vdash k_1, \dots, \lambda_a \vdash k_a$.

4.2f. **Theorem.** (i) *The functor $R^{G_n}_{G_{d\nu}}$ sends $C_{\sigma,\nu}$-modules with Δ-filtrations (resp. ∇-filtrations) to $C_{\sigma,k}$-modules with Δ-filtrations (resp. ∇-filtrations).*

(ii) *The functor ${}^*R^{G_n}_{G_{d\nu}}$ sends $C_{\sigma,k}$-modules with Δ-filtrations (resp. ∇-filtrations) to $C_{\sigma,\nu}$-modules with Δ-filtrations (resp. ∇-filtrations).*

Proof. (i) We first need to observe that the functor $R^{G_n}_{G_{d\nu}}$ sends $C_{\sigma,\nu}$-modules to $C_{\sigma,k}$-modules. It suffices to check this on projective $C_{\sigma,\nu}$-modules. In turn, since according to the Levi analogue of Theorem 3.4g every projective $C_{\sigma,\nu}$-module is a submodule of $M^\nu(\sigma)$, we just need to check that $R^{G_n}_{G_{d\nu}} M^\nu(\sigma)$ is a $C_{\sigma,k}$-module. But this is clear since by transitivity of Harish-Chandra induction, $R^{G_n}_{G_{d\nu}} M^\nu(\sigma) \cong M^k(\sigma)$. So it makes sense to regard the functor $R^{G_n}_{G_{d\nu}}$ as a functor from $\mathrm{mod}(C_{\sigma,\nu})$ to $\mathrm{mod}(C_{\sigma,k})$.

Now to show that $R^{G_n}_{G_{d\nu}}$ sends modules with Δ-filtrations to modules with Δ-filtrations, we just need to check by exactness that

$$R^{G_n}_{G_{d\nu}}(\Delta(\sigma,\lambda_1) \boxtimes \cdots \boxtimes \Delta(\sigma,\lambda_a))$$

has a Δ-filtration, for any $\lambda_1 \vdash k_1, \dots, \lambda_a \vdash k_a$. According to Theorem 4.2a,

$$R^{G_n}_{G_{d\nu}}(\Delta(\sigma,\lambda_1) \boxtimes \cdots \boxtimes \Delta(\sigma,\lambda_a)) \cong \beta_{\sigma,h,k}(\Delta_h(\lambda_1') \otimes \cdots \otimes \Delta_h(\lambda_a')).$$

So the result follows since $\Delta_h(\lambda_1') \otimes \cdots \otimes \Delta_h(\lambda_a')$ has a Δ-filtration as an $S_{h,k}$-module thanks to (1.3c).

This proves (i) in the case of Δ-filtrations, and the result for ∇-filtrations follows immediately on taking duals, since contravariant duality commutes with Harish-Chandra induction.

(ii) Again we first check that ${}^*R^{G_n}_{G_{d\nu}}$ sends $C_{\sigma,k}$-modules to $C_{\sigma,\nu}$-modules. Making the same reductions as before, we need to observe that ${}^*R^{G_n}_{G_{d\nu}} M^k(\sigma)$ is a $C_{\sigma,\nu}$-module, which we checked in Lemma 3.2f. So it makes sense to regard the functor ${}^*R^{G_n}_{G_{d\nu}}$ as a functor from $\mathrm{mod}(C_{\sigma,k})$ to $\mathrm{mod}(C_{\sigma,\nu})$.

Now we prove (ii) in the case of ∇-filtrations; the analogous result for Δ-filtrations follows on dualizing as before. So take $N \in \mathrm{mod}(C_{\sigma,k})$ with a ∇-filtration. Using the cohomological criterion for ∇-filtrations [Do7, A2.2(iii)], we need to show that

$$\mathrm{Ext}^1_{C_{\sigma,\nu}}(M, {}^*R^{G_n}_{G_{d\nu}} N) = 0$$

for all $M \in \mathrm{mod}(C_{\sigma,\nu})$ with a Δ-filtration. So take such a module M. By (i) and the cohomological criterion for ∇-filtrations, we know that

$$\mathrm{Ext}^1_{C_{\sigma,k}}(R^{G_n}_{G_{d\nu}} M, N) = 0.$$

So the result will follow if we can prove:

(4.2g) *For all $M \in \mathrm{mod}(C_{\sigma,\nu}), N \in \mathrm{mod}(C_{\sigma,k})$ and $i \geq 0$,*

$$\mathrm{Ext}^i_{C_{\sigma,\nu}}(M, {}^*R^{G_n}_{G_{d\nu}} N) \cong \mathrm{Ext}^i_{C_{\sigma,k}}(R^{G_n}_{G_{d\nu}} M, N).$$

Fix $M \in \text{mod}(C_{\sigma,\nu})$. The adjoint functor property gives us an isomorphism of functors $\text{Hom}_{C_{\sigma,\nu}}(M,?) \circ {}^*R^{G_n}_{G_{d\nu}} \cong \text{Hom}_{C_{\sigma,k}}(R^{G_n}_{G_{d\nu}}M,?)$. Since ${}^*R^{G_n}_{G_{d\nu}}$ is exact and sends injectives to injectives (being adjoint to the exact functor $R^{G_n}_{G_{d\nu}}$), a standard degree shifting argument (cf. [Ja, I.4.1(3)]) now gives (4.2g). □

4.2h. **Corollary.** *Take $\lambda \vdash k$ and partitions $\lambda_1 \vdash k_1, \dots, \lambda_a \vdash k_a$. Then, both of*

(i) *the multiplicity of $\Delta(\sigma,\lambda)$ in a Δ-filtration of $R^{G_n}_{G_{d\nu}} \Delta(\sigma,\lambda_1) \boxtimes \cdots \boxtimes \Delta(\sigma,\lambda_a)$,*

(ii) *the multiplicity of $\Delta(\sigma,\lambda_1) \boxtimes \cdots \boxtimes \Delta(\sigma,\lambda_a)$ in a Δ-filtration of ${}^*R^{G_n}_{G_{d\nu}} \Delta(\sigma,\lambda)$,*

are the same as in characteristic zero, i.e. are given by the Littlewood-Richardson rule.

Proof. The fact that the modules in (i) and (ii) have Δ-filtrations follows from Theorem 4.2f. That the multiplicities are the same as in characteristic zero follows in case (i) from the analogous well-known fact about tensor product multiplicities over the q^d-Schur algebra, using Theorem 4.2a. The conclusion in case (ii) follows from (i) and adjointness, together with the usual properties of Δ- and ∇-filtrations. □

4.3. p-Singular classes

Throughout the section fix a p-singular element $\tau \in \bar{\mathbb{F}}_q^\times$ of degree e over $\mathbb{F}_q$. Let σ be the p-regular part of τ, of degree d over $\mathbb{F}_q$. By (2.1a), $e = dm$ where $m = \ell(d)p^r$ for some $r \geq 0$. Choose $l \geq 1$, set $k = lm$ and let $n = kd = le$.

Fixing an integer $h \geq k$, let $S_{h,k}$ (resp. $\bar{S}_{h,l}$) denote the q^d-Schur algebra $S_{F,q^d}(h,k)$ (resp. the q^e-Schur algebra $S_{F,q^e}(h,l)$). We observe that the image of q^e in F is 1, so in fact $\bar{S}_{h,l}$ is just the classical Schur algebra $S_{F,1}(h,l)$. Also let A_h (resp. $\bar{A}_h$) denote the quantized coordinate ring $A_{F,v^d}(h)$ (resp. $A_{F,1}(h)$), choosing a square root v^d of q^d in F in the same way as in §1.3. Let V_h (resp. $\bar{V}_h$) denote the natural h-dimensional A_h-comodule (resp. the natural $\bar{A}_h$-comodule).

Recall that for a right $\bar{A}_h$-comodule M, we have defined its rth Frobenius twist $M^{[r]}$, which is a right A_h-comodule, by inflation along the bialgebra map $F_r : \bar{A}_h \to A_h$ of (1.3.5). If M is an $\bar{S}_{h,l}$-module, then the Frobenius twist $M^{[r]}$ is an $S_{h,k}$-module. Since F_r is a bialgebra map, the operation of taking Frobenius twists commutes with tensor products.

Now, σ is p-regular, so the theory of the previous sections holds for σ, thanks to Lemma 2.4c and Lemma 2.5f. Let $\beta_{\sigma,h,k} : \text{mod}(S_{h,k}) \to \text{mod}(C_{\sigma,k})$ denote the equivalence of categories of (3.5.2). Our first goal is to extend our theory to the p-singular element τ.

4.3a. **Lemma.** *The Brauer character of $L(\sigma,(l^m))$ agrees with the restriction of $\chi_{\tau,(l)}$ to p-regular classes of G_n.*

Proof. Recall that $\beta_{\sigma,h,k}(L_h(m^l)) = L(\sigma,(l^m))$ by (3.5.3). Applying the Morita equivalence $\beta_{\sigma,h,k}$ to Lemma 1.3f, using Theorem 4.1c, we see that the Brauer character of $L(\sigma, (l^m))$ is equal to the restriction of the generalized character

$$\frac{(-1)^{k+l}}{l!} \sum_{\lambda \vdash l} \sum_{\mu \vdash k} c_\lambda \phi_\mu(m\lambda) \chi_{\sigma,\mu}$$

to p-regular classes. So the lemma follows from Lemma 2.3c. □

4.3b. **Theorem.** *For any $l \geq 1$, there is an isomorphism of FG_n-modules:*

$$M^l(\tau) \cong \beta_{\sigma,h,k}(\underbrace{\bar{V}_h^{[r]} \otimes \cdots \otimes \bar{V}_h^{[r]}}_{l \text{ times}}).$$

Hence, $M(\tau)$ is an irreducible FG_e-module and, for any $l \geq 1$, $M^l(\tau)$ has precisely Par_l *non-isomorphic composition factors.*

Proof. Consider first the case that $l = 1$. By Lemma 4.3a, $M(\tau)$ has the same Brauer character as $L(\sigma, (1^m))$. Since the latter is an irreducible module, this implies immediately that $M(\tau) \cong L(\sigma, (1^m))$, so that $M(\tau)$ is also irreducible. Since $L(\sigma, (1^m)) = \beta_{\sigma,h,k}(L_h((m))) = \beta_{\sigma,h,k}(\bar{V}_h^{[r]})$, this proves the result in the case $l = 1$.

Now for $l > 1$, the isomorphism $M^l(\tau) \cong \beta_{\sigma,h,k}(\bar{V}_h^{[r]} \otimes \cdots \otimes \bar{V}_h^{[r]})$ is immediate using Theorem 4.2a and the definition of $M^l(\tau)$.

Finally, to see that $M^l(\tau)$ has precisely Par_l non-isomorphic composition factors, observe that

$$\underbrace{\bar{V}_h^{[r]} \otimes \cdots \otimes \bar{V}_h^{[r]}}_{l \text{ times}} \cong (\underbrace{\bar{V}_h \otimes \cdots \otimes \bar{V}_h}_{l \text{ times}})^{[r]}.$$

This has precisely Par_l non-isomorphic composition factors as an $S_{h,k}$-module, as the untwisted tensor space $\bar{V}_h^{\otimes l}$ has precisely Par_l non-isomorphic composition factors as an $\bar{S}_{h,l}$-module. □

Theorem 4.3b (combined with Lemma 2.4c and Lemma 2.5f) verifies the assumptions (A1) and (A2) from §3.2 for *every* $\sigma \in \bar{\mathbb{F}}_q$. So the main Morita theorem of (3.5a), and the subsequent results obtained so far, are true for general σ, hence for our fixed p-singular element τ. In particular, we associate to τ the (p-singular) cuspidal algebra $C_{\tau,l}$, defined as in (3.2.1). By the analogue of (3.5a), there is an equivalence of categories

$$\beta_{\tau,h,l} : \mathrm{mod}(\bar{S}_{h,l}) \to \mathrm{mod}(C_{\tau,l}),$$

defined as in (3.5.2). We have now recovered the main results from the paper of James [J$_2$] with our approach.

4.3c. **Remarks.** (i) One can also consider an analogue of the cuspidal algebra for an arbitrary block-diagonal element $s \in G_n$, namely, the quotient algebra

$$C_s = C_{F,s}(GL_n(\mathbb{F}_q)) = FG_n / \mathrm{ann}_{FG_n}(M(s))$$

where $M(s)$ is as in (2.4.3). If s is p-regular, the Cline-Parshall-Scott theorem mentioned in Remark 4.1e gives an analogue of the Morita theorem (even over $\mathcal{O}$) for the algebra C_s: the algebra C_s is Morita equivalent to $\bigotimes_{i=1}^a S_{F,q^{d_i}}(h_i, k_i)$ for integers $h_i \geq k_i$, where s is of the form (2.1.1). The same is true for somewhat more general s, namely, if s is "reduction stable" (see [DF] or [D$_4$, 4.26]), but certainly false in general.

(ii) Theorem 4.3b also yields an alternative proof of [D$_2$, 3.5]: for an arbitrary block-diagonal element $s \in G_n$, the associated cuspidal $G_{\delta(s)}$-module of (2.4.4) remains irreducible modulo p, since it is an outer tensor product of factors all of which remain irreducible.

We now associate to τ the $C_{\tau,l}$- (hence FG_n-) modules $L(\tau, \lambda), \Delta(\tau, \lambda)$ and $\nabla(\tau, \lambda)$ for each $\lambda \vdash l$, defined as in (3.5.3)–(3.5.5) but using the functor $\beta_{\tau,h,l}$ instead. The next result realizes these modules alternatively as modules in the category $\mathrm{mod}(C_{\sigma,k})$ (for (iv), see also [DJ$_2$, [Lemma 2.3]).

4.3d. **Theorem.** *For any $\nu \models l$ and $\lambda \vdash l$, we have the following FG_n-module isomorphisms:*

(i) $\dot{\Lambda}^\nu(\tau) \cong \beta_{\sigma,h,k}(\Lambda^\nu(\bar{V}_h)^{[r]})$;

(ii) $\dot{Z}^\nu(\tau) \cong \beta_{\sigma,h,k}(Z^\nu(\bar{V}_h)^{[r]})$;

(iii) $\Delta(\tau, \lambda) \cong \beta_{\sigma,h,k}(\bar{\Delta}_l(\lambda')^{[r]})$;

(iv) $L(\tau, \lambda) \cong \beta_{\sigma,h,k}(\bar{L}_l(\lambda')^{[r]}) \cong L(\sigma, (m\lambda')')$.

Proof. (i) Since $\dot{\Lambda}^\nu(\tau) \cong R_{G_{e\nu}}^{G_n} \Lambda^\nu(\tau)$, it suffices applying Theorem 4.2a to prove this in the special case $\nu = (l)$. Recall (e.g. by Theorem 3.5e) that $\Lambda^l(\tau)$ is isomorphic to the module $\Delta(\tau, (l))$. So by Theorem 4.1c, the Brauer character of $\Lambda^l(\tau)$ is equal to the restriction of $\chi_{\tau,(l)}$ to p-regular classes, which by Lemma 4.3a is the same as the Brauer character of $L(\sigma, (l^m))$. Since the latter is an irreducible module, we deduce that

$$\Lambda^l(\tau) \cong L(\sigma, (l^m)) = \beta_{\sigma,h,k}(L_h((m^l))) \cong \beta_{\sigma,h,k}(\bar{L}_h((1^l))^{[r]}) \cong \beta_{\sigma,h,k}(\Lambda^l(\bar{V}_h)^{[r]}).$$

(ii) Again, we just need to prove this in the special case $\nu = (l)$. So, we need to prove that

$$Z^l(\tau) \cong \beta_{\sigma,h,k}(Z^l(\bar{V}_h)^{[r]}).$$

We first observe that $Z^l(\bar{V}_h)^{[r]}$ is a quotient of $Z^k(V_h)$ as $S_{h,k}$-modules. This follows from the universal property of standard modules [PW, (8.10.2)]: $Z^l(\bar{V}_h)^{[r]} \cong \bar{\Delta}_h((l))^{[r]}$ is generated by a highest weight vector of weight (k) so is a quotient of the universal highest weight module $Z^k(V_h) \cong \Delta_h((k))$. Now, $Z^l(\bar{V}_h)^{[r]}$ is a submodule of $(\bar{V}_h^{\otimes l})^{[r]}$. So we see that there is a non-zero homomorphism

$$Z^k(V_h) \to (\bar{V}_h^{\otimes l})^{[r]}$$

with image $Z^l(\bar{V}_h)^{[r]}$. Now apply the Morita equivalence $\beta_{\sigma,h,k}$ and Lemma 4.3a to deduce that there is a non-zero homomorphism

$$Z^k(\sigma) \to M^l(\tau)$$

with image $\beta_{\sigma,h,k}(Z^l(\bar{V}_h)^{[r]})$. Now, $Z^k(\sigma)$ is a quotient of Γ_n, so we have shown that the image of some non-zero homomorphism

$$\Gamma_n \to M^l(\tau)$$

is isomorphic to $\beta_{\sigma,h,k}(Z^l(\bar{V}_h)^{[r]})$. But by (3.4h), the image of any such homomorphism is precisely the submodule $Z^l(\tau)$ of $M^l(\tau)$.

(iii) By Theorem 3.5e(i), $\Delta(\tau, \lambda)$ is isomorphic to the image of any non-zero homomorphism from $\dot{Z}^{\lambda'}(\tau)$ to $\dot{\Lambda}^{\lambda}(\tau)$. So using (i) and (ii), $\Delta(\tau, \lambda)$ is isomorphic to the image of any non-zero homomorphism

$$\beta_{\sigma,h,k}(Z^{\lambda'}(\bar{V}_h)^{[r]}) \to \beta_{\sigma,h,k}(\Lambda^{\lambda}(\bar{V}_h)^{[r]}).$$

So using the Morita equivalence, we just need to check that the image of any non-zero homomorphism

$$Z^{\lambda'}(\bar{V}_h)^{[r]} \to \Lambda^{\lambda}(\bar{V}_h)^{[r]}$$

is isomorphic to $\bar{\Delta}_h(\lambda')^{[r]}$. But by (1.3d), the image of any non-zero homomorphism

$$Z^{\lambda'}(\bar{V}_h) \to \Lambda^{\lambda}(\bar{V}_h)$$

is isomorphic to $\bar{\Delta}_h(\lambda')$, so the conclusion follows on taking Frobenius twists.

(iv) The simple head of $\bar{\Delta}_h(\lambda')$ is isomorphic to $\bar{L}_h(\lambda')$. So the simple head of $\bar{\Delta}_h(\lambda')^{[r]}$ is isomorphic to $\bar{L}_h(\lambda')^{[r]}$. Now applying (iii), we see that $\beta_{\sigma,h,k}(\bar{L}_h(\lambda')^{[r]})$ is isomorphic to the simple head of $\Delta(\tau, \lambda)$, namely, $L(\tau, \lambda)$. This proves the first isomorphism. For the second isomorphism, note that

$$\beta_{\sigma,h,k}(\bar{L}_h(\lambda')^{[r]}) \cong \beta_{\sigma,h,k}(L_h(m\lambda')) = L(\sigma, (m\lambda')')$$

using a special case of (1.3e). □

Now we can deduce the non-defining characteristic analogue of Steinberg's tensor product theorem (cf. [DDu2]).

4.3e. **Theorem.** *Suppose that $\sigma \in \bar{\mathbb{F}}_q^\times$ is a p-regular element of degree d over $\mathbb{F}_q$. Let λ be a partition of k and let $\lambda_{-1}, \lambda_0, \dots, \lambda_a$ be the partitions such that*

$$\lambda' = \lambda'_{-1} + \ell(d)\lambda'_0 + \ell(d)p\lambda'_1 + \cdots + \ell(d)p^a\lambda'_a$$

is the $(\ell(d), p)$-adic expansion of λ' (the λ_i are uniquely determined as $h \geq k$). For each $i = 0, \dots, a$, choose a p-singular element $\sigma_i \in \bar{\mathbb{F}}_q^\times$ of degree $d\ell(d)p^i$ over $\mathbb{F}_q$ with p-regular part conjugate to σ (such elements exist by (2.1a)). Then, $L(\sigma, \lambda)$ is isomorphic to the module obtained by Harish-Chandra induction from

$$L(\sigma, \lambda_{-1}) \boxtimes L(\sigma_0, \lambda_0) \boxtimes L(\sigma_1, \lambda_1) \boxtimes \cdots \boxtimes L(\sigma_a, \lambda_a).$$

Proof. Because of Theorem 4.2a and Theorem 4.3d(iv), the theorem is a direct restatement of (1.3e) under the Morita equivalence. □

4.3f. **Remark.** There are two different parametrizations of the irreducible FG_n-modules that appear in the literature. The first arises from Harish-Chandra theory and was introduced in [D1, D2]. It involves pairs $(s, \underline{\lambda})$, where s runs over all p-regular semisimple classes of G_n together with certain p-singular classes, and $\underline{\lambda}$ runs over *p-regular* multipartitions of $\kappa(s)$. The second parametrization is described in the next section and was introduced by James in [J2]. It involves pairs $(s, \underline{\lambda})$ where s runs over representatives for the p-regular semisimple classes *only* but now $\underline{\lambda}$ runs over *all* multipartitions of $\kappa(s)$. The combinatorics to translate between the two parametrizations was explained originally in [DJ2]. Theorem 4.3e can be used to give an alternative, more representation theoretic proof of main result of *loc. cit.*. We refer the reader to [DDu2, D4] for further details of these matters.

4.4. Blocks and decomposition numbers

Let us now introduce notation for irreducible and standard modules of FG_n associated to an *arbitrary* semisimple element. So suppose that s is a block-diagonal element of G_n written in the form (2.1.1). For $\underline{\lambda} = (\lambda_1, \dots, \lambda_a) \vdash \kappa(s)$ and R equal to one of K, F or $\mathcal{O}$, define the *standard FG_n-module*:

$$\Delta(s, \underline{\lambda})_R = R^{G_n}_{G_{\pi(s)}}(\Delta(\sigma_1, \lambda_1)_R \boxtimes \cdots \boxtimes \Delta(\sigma_a, \lambda_a)_R)$$

We will write simply $\Delta(s, \underline{\lambda})$ for $\Delta(s, \underline{\lambda})_F$ over F. Since Harish-Chandra induction commutes with base change, we have immediately by Theorem 4.1c and (2.3e):

(4.4a) *$\Delta(s, \underline{\lambda})_{\mathcal{O}}$ is $\mathcal{O}$-free of finite rank with*

$$\Delta(s, \underline{\lambda})_K \cong K \otimes_{\mathcal{O}} \Delta(s, \underline{\lambda})_{\mathcal{O}},$$
$$\Delta(s, \underline{\lambda})_F \cong F \otimes_{\mathcal{O}} \Delta(s, \underline{\lambda})_{\mathcal{O}}.$$

Moreover, the character of $\Delta(s, \underline{\lambda})_K$ is $\chi_{s,\underline{\lambda}}$, so $\Delta(s, \underline{\lambda})$ is a p-modular reduction of a KG_n-module affording the character $\chi_{s,\underline{\lambda}}$.

To describe *all* irreducible FG_n-modules, suppose now in addition that s is p-regular. Define

$$L(s, \underline{\lambda}) = R^{G_n}_{G_{\pi(s)}}(L(\sigma_1, \lambda_1) \boxtimes \cdots \boxtimes L(\sigma_a, \lambda_a)). \tag{4.4.1}$$

The FG_{n_i}-module $L(\sigma_i, \lambda_i)$ is an irreducible module for the block B_{σ_i,k_1}, and is isomorphic to the simple head of $\Delta(\sigma_i, \lambda_i)$. So, in the notation of Theorem 2.4e, $L(\sigma_1, \lambda_1) \boxtimes \cdots \boxtimes L(\sigma_a, \lambda_a)$ is an irreducible module for B_s^{Levi}, and is isomorphic the simple head of $\Delta(\sigma_1, \lambda_1) \boxtimes \cdots \boxtimes \Delta(\sigma_a, \lambda_a)$. Now Theorem 2.4e immediately gives that $L(s, \underline{\lambda})$ is irreducible and is isomorphic to the simple head of $\Delta(s, \underline{\lambda})$. Moreover, recalling (2.4.6) we have:

(4.4b) *For p-regular s, the modules* $\{L(s,\underline{\lambda}) \mid \underline{\lambda} \vdash \kappa(s)\}$ *give a complete set of non-isomorphic irreducible* B_s*-modules. So, the modules* $\{L(s,\underline{\lambda}) \mid s \in \mathcal{C}_{ss,p'}, \underline{\lambda} \vdash \kappa(s)\}$ *give a complete set of non-isomorphic irreducible* FG_n*-modules.*

Now we discuss two closely related problems. First, we would like to describe the *p-blocks* of the group G_n. So we would like to know precisely when the ordinary irreducible characters $\chi_{s,\underline{\lambda}}$ and $\chi_{t,\underline{\mu}}$ belong to the same p-block, for any $s, t \in \mathcal{C}_{ss}, \underline{\lambda} \vdash \kappa(s), \underline{\mu} \vdash \kappa(t)$. In view of (4.4a), we can equivalently consider when the modules $\Delta(s,\underline{\lambda})$ and $\Delta(t,\underline{\mu})$ belong to the same block of the algebra FG_n. Second, we would like to say something about the *p-modular decomposition numbers* of G_n, that is, composition multiplicities of the form $[\Delta(s,\underline{\lambda}) : L(t,\underline{\mu})]$ for any $s \in \mathcal{C}_{ss}, t \in \mathcal{C}_{ss,p'}, \underline{\lambda} \vdash \kappa(s), \underline{\mu} \vdash \kappa(t)$.

We can make some basic reductions to both of these problems which will simplify notation considerably. First, we apply (2.4.6) to see that we just need to describe the p-blocks and the decomposition numbers of the algebra B_t, for some fixed $t \in \mathcal{C}_{ss.p'}$. Then we can apply the Morita equivalence of Theorem 2.4e to reduce further to the special case that $t = (\sigma)^k$ for some p-regular $\sigma \in \bar{\mathbb{F}}_q^\times$ of degree d over $\mathbb{F}_q$, where $n = kd$. In other words, we just need to describe the blocks and the decomposition numbers of the algebra $B_{\sigma,k}$.

So fix now a p-regular $\sigma \in \bar{\mathbb{F}}_q^\times$ and $h \geq k \geq 1$. We recall from (2.4b) that the standard $B_{\sigma,k}$-modules are precisely the modules $\Delta(s,\underline{\lambda})$ (corresponding to the characters $\chi_{s,\underline{\lambda}}$) for $\underline{\lambda} \vdash \kappa(s)$ and $s \in \mathcal{C}_{ss}$ of the form (2.1.1) with the p-regular part of each σ_i conjugate to σ. Our first lemma relates this module $\Delta(s,\underline{\lambda})$ to the q^d-Schur algebra $S_{h,k} = S_{F,q^d}(h,k)$.

4.4c. **Lemma.** *For any block-diagonal element* $s = (\sigma_1)^{k_1} \dots (\sigma_a)^{k_a}$ *of the form (2.1.1), with the p-regular part of each* σ_i *conjugate to* σ*, and any* $\underline{\lambda} = (\lambda_1, \dots, \lambda_a) \vdash \kappa(s)$,

$$\Delta(s,\underline{\lambda}) \cong \beta_{\sigma,h,k}\left(\Delta_h(\lambda_1')^{[r_1]} \otimes \cdots \otimes \Delta_h(\lambda_a')^{[r_a]}\right)$$

where each r_i *is determined by the equation* $\deg(\sigma_i) = d\ell(d)p^{r_i}$.

Proof. By definition, $\Delta(\sigma,\underline{\lambda})$ is the FG_n-module obtained by Harish-Chandra induction from the $FG_{\pi(s)}$-module $\Delta(\sigma_1,\lambda_1) \boxtimes \cdots \boxtimes \Delta(\sigma_a,\lambda_a)$. By Theorem 4.3d(iii), each $\Delta(\sigma_i,\lambda_i)$ is isomorphic to $\beta_{\sigma,h,k_i}(\Delta_h(\lambda_i')^{[r_i]})$. The lemma now follows on applying Theorem 4.2a. □

Now we can explain the algorithm in [DJ3, §7] showing how to calculate the p-modular decomposition numbers of G_n from knowledge of the decomposition numbers of quantum linear groups. Actually, we will give a somewhat more precise formula relating the two. We need a little notation for the statement: for $\lambda, \mu, \nu_1, \dots, \nu_a \in \Lambda^+(h)$ and $e \geq 1$, $\mathrm{D}^e_{\lambda,\mu}$ denotes the decomposition number $[\Delta_h(\lambda) : L_h(\mu)]$ of the quantized coordinate ring $A_{F,q^e}(h)$ and $\mathrm{LR}^e_{\nu_1,\dots,\nu_a;\mu}$ denotes the tensor product composition multiplicity $[L_h(\nu_1) \otimes \cdots \otimes L_h(\nu_a) : L_h(\mu)]$ for $A_{F,q^e}(h)$.

4.4d. **Theorem.** *Suppose that $s = (\sigma_1)^{k_1} \dots (\sigma_a)^{k_a}$ is a block-diagonal element of G_n of the form (2.1.1), with each σ_i of degree $d_i = m_i d$ over $\mathbb{F}_q$ having p-regular part conjugate to σ. Then, for any $\underline{\lambda} = (\lambda_1, \dots, \lambda_a) \vdash \kappa(s)$ and any $\mu \vdash k$, we have the equality:*

$$[\Delta(s, \underline{\lambda}) : L(\sigma, \mu)] = \sum_{\nu_1 \vdash k_1} \sum_{\nu_2 \vdash k_2} \cdots \sum_{\nu_a \vdash k_a} \left[\prod_{i=1}^{a} \mathrm{D}^{d_i}_{\lambda_i', \nu_i'} \right] \mathrm{LR}^d_{m_1\nu_1', \dots, m_a\nu_a'; \mu'} \, .$$

Proof. Using Lemma 4.4c and the Morita equivalence, we need to prove that for any $\lambda_i \in \Lambda^+(h, k_i)$ and $\mu \in \Lambda^+(h, k)$,

$$[\bar{\Delta}_h(\lambda_1)^{[r_1]} \otimes \cdots \otimes \bar{\Delta}_h(\lambda_a)^{[r_a]} : L_h(\mu)] = \sum_{\nu_1 \vdash k_1} \cdots \sum_{\nu_a \vdash k_a} \left[\prod_{i=1}^{a} \mathrm{D}^{d_i}_{\lambda_i, \nu_i} \right] \mathrm{LR}^d_{m_1\nu_1, \dots, m_a\nu_a; \mu}$$

where the r_i are defined by $m_i = \ell(d)p^{r_i}$. For $\nu \in \Lambda^+(h, m_i k_i)$, the composition multiplicity $[\bar{\Delta}_h(\lambda_1)^{[r_1]} : L_h(\nu)]$ is zero unless $\nu = m_i \nu_i$ for some $\nu_i \in \Lambda^+(h, k_i)$, when it equals $\mathrm{D}^{d_i}_{\lambda_i, \nu_i}$. It follows that $\bar{\Delta}_h(\lambda_1)^{[r_1]} \otimes \cdots \otimes \bar{\Delta}_h(\lambda_a)^{[r_a]}$ has the same composition factors as

$$\bigoplus_{\nu_1 \vdash k_1} \cdots \bigoplus_{\nu_a \vdash k_a} \left[\prod_{i=1}^{a} \mathrm{D}^{d_i}_{\lambda_i, \nu_i} \right] L_h(m_1\nu_1) \otimes \cdots \otimes L_h(m_a\nu_a).$$

Now the theorem follows. □

Now consider the block theory. We first recall the description of the blocks of the q^d-Schur algebra $S_{h,k}$ from [JM, Theorem 4.24]. For $s \geq 1$, recall that the *s-core* of a partition λ is the partition obtained by successively removing as many rim hooks of length s as possible. Then, for $\lambda, \mu \in \Lambda^+(h, k)$ (and $h \geq k$ as always), the irreducible $S_{h,k}$-modules $L_h(\lambda)$ and $L_h(\mu)$ belong to the same block if and only if either:

(i) $\ell(d) = 1$ and λ and μ have the same p-core;

(ii) $\ell(d) > 1$ and λ and μ have the same $\ell(d)$-core.

Now, since $C_{\sigma,k}$ is Morita equivalent to $S_{h,k}$ (and the cores of λ and λ' are always transpose to one another), this combinatorics also determines when the modules $L(\sigma, \lambda)$ and $L(\sigma, \mu)$ belong to the same block of the cuspidal algebra $C_{\sigma,k}$. Recalling that $C_{\sigma,k}$ is a quotient of the algebra $B_{\sigma,k}$ and that the two algebras have the same irreducible modules, $B_{\sigma,k}$ can in general have fewer blocks than $C_{\sigma,k}$.

4.4e. **Lemma.** *If $\ell(d) = 1$ (i.e. $q^d \equiv 1 \pmod p$), then the algebra $B_{\sigma,k}$ has just one block.*

Proof. Let $\lambda = (l_1, \dots, l_a)$ and $\mu = (l_1 + l_a, l_2, \dots, l_{a-1})$ be partitions of k. Following the strategy of [DJ3, Lemma 7.10], we show that $L(\sigma, \lambda)$ and $L(\sigma, \mu)$ are linked. This is enough to complete the proof, for then by induction on a, we see that $L(\sigma, \lambda)$ belongs to the same block as $L(\sigma, (k))$, whence all $L(\sigma, \lambda)$ belong to the same block.

Since $\ell(d) = 1$, we can by (2.1a) choose some p-singular $\tau \in \bar{\mathbb{F}}_q^\times$ of degree d over $\mathbb{F}_q$ with p-regular part conjugate to σ. Writing $\nu = (l_1, \dots, l_{a-1})$, the FG_n-module M obtained from $\Delta(\sigma, \nu) \boxtimes \Delta(\tau, (l_a))$ by Harish-Chandra induction is a standard module for $B_{\sigma,k}$. So to prove that $L(\sigma, \lambda)$ and $L(\sigma, \mu)$ are linked, it suffices to show that they are both composition factors of M.

Equivalently, thanks to Lemma 4.4c and the Morita equivalence, we need to show that the $S_{h,k}$-module $\Delta_h(\nu') \otimes \Delta_h((1^{l_a}))$ contains both $L_h(\lambda')$ and $L_h(\mu')$ as composition factors. But now $\Delta_h(\nu') \otimes \Delta_h((1^{l_a}))$ has a Δ-filtration, and a calculation involving the Littlewood-Richardson rule shows that this filtration has factors isomorphic to $\Delta_h(\lambda')$ and to $\Delta_h(\mu')$. Hence, it certainly has composition factors $L_h(\lambda')$ and $L_h(\mu')$ as desired. □

4.4f. **Lemma.** *Suppose that $s = (\sigma_1)^{k_1} \dots (\sigma_a)^{k_a}$ is of the form (2.1.1), with σ_1 conjugate to σ, and all other σ_i having p-regular part conjugate to σ. If $L(\sigma, \mu)$ is a composition factor of $\Delta(s, \underline{\lambda})$, for some $\underline{\lambda} = (\lambda_1, \dots, \lambda_a) \vdash \kappa(s)$ and $\mu \vdash k$, then μ and λ_1 have the same $\ell(d)$-cores.*

Proof. If $\ell(d) = 1$, the lemma is trivially true. So assume that $\ell(d) > 1$. Then by (2.1a), all of $\sigma_1, \dots, \sigma_a$ are of degree strictly greater than d. So applying Lemma 4.4c, we see that $\Delta(s, \underline{\lambda})$ corresponds under the Morita equivalence to an $S_{h,k}$-module of the form $\Delta_h(\lambda_1') \otimes M$ where M is a pure Frobenius twist. Now it suffices to prove that $L_h(\mu')$ is a composition factor of $\Delta_h(\lambda_1') \otimes M$ only if μ' and λ_1' have the same $\ell(d)$-cores.

Using Steinberg's tensor product theorem (1.3e) and the known block theory of $S_{h,k}$, all composition factors of $\Delta_h(\lambda_1')$ are of the form $L_h(\nu) \otimes N$ where ν is $\ell(d)$-restricted and has the same $\ell(d)$-core as λ_1', and N is a pure Frobenius twist. Hence, using the tensor product theorem once more, all composition factors of $\Delta_h(\lambda_1') \otimes M$ are of the form $L_h(\nu) \otimes \bar{L}_h(\gamma)^{[0]}$ for such ν. So, if $L_h(\mu')$ is a composition factor of $\Delta_h(\lambda_1') \otimes M$ then $\mu' = \nu + \ell(d)\gamma$, which has the same $\ell(d)$-core as ν hence as λ_1'. □

Finally, we can determine the blocks of $B_{\sigma,k}$. The following theorem is equivalent, after making the elementary reductions described above, to Fong and Srinivasan's theorem [FS, (7A)].

4.4g. **Theorem.** *The blocks of the algebra $B_{\sigma,k}$ are parametrized by the set of $\ell(d)$-cores of partitions of k. Moreover, given $s = (\sigma_1)^{k_1} \dots (\sigma_a)^{k_a}$ of the form (2.1.1), with σ_1 conjugate to σ and all other σ_i having p-regular part conjugate to σ, and $\underline{\lambda} = (\lambda_1, \dots, \lambda_a) \vdash \kappa(s)$, the standard module $\Delta(s, \underline{\lambda})$ belongs to the block parametrized by the $\ell(d)$-core of λ_1.*

Proof. If $\ell(d) = 1$, this is immediate from Lemma 4.4e. For $\ell(d) > 1$, use Lemma 4.4f and the block theory of $S_{h,k}$ as described above. □

4.5. The Ringel dual of the cuspidal algebra

If S is any quasi-hereditary algebra over F with weight poset $(\Lambda^+, \leq)$, an S-module M has a Δ*-filtration* (resp. a ∇*-filtration*) if it has a filtration with sections isomorphic to modules of the form $\Delta(\lambda)$ (resp. $\nabla(\lambda)$), $\lambda \in \Lambda^+$. Recall that in any such Δ-filtration of M, the multiplicity of a particular $\Delta(\lambda)$ for fixed $\lambda \in \Lambda^+$ us uniquely determined; we write $[M : \Delta(\lambda)]_\Delta$ for this multiplicity. Say that an S-module is *tilting* if it has both a Δ-filtration and a ∇-filtration. Ringel [R] has shown that for each $\lambda \in \Lambda^+$, there is a unique indecomposable tilting module $T(\lambda)$ such that $[T(\lambda) : \Delta(\lambda)]_\Delta = 1$ and, for $\mu \in \Lambda^+$, $[T(\lambda) : \Delta(\mu)]_\Delta = 0$ unless $\mu \leq \lambda$. Furthermore, every tilting module is isomorphic to a direct sum of these indecomposable tilting modules modules $T(\lambda)$, $\lambda \in \Lambda^+$.

Following the language of [Do$_7$, Appendix], a *full tilting module* is a tilting module that contains every $T(\lambda)$, $\lambda \in \Lambda^+$ as a summand with non-zero multiplicity. Given such a full tilting module T, the *Ringel dual* of S relative to T is the algebra $S^\star = \mathrm{End}_S(T)^{\mathrm{op}}$. Here, we are writing endomorphisms on the right, so T is naturally a right $\mathrm{End}_S(T)$-module, hence a left $S^\star$-module. Ringel [R] showed that $S^\star$ is also a quasi-hereditary algebra with weight poset Λ^+, but ordered with the opposite order to the original partial ordering on Λ^+. We briefly indicate one approach to the proof of this, since we will need the notation shortly. We define the contravariant functor

$$\gamma : \mathrm{mod}(S) \to \mathrm{mod}(S^\star), \qquad \gamma = \mathrm{Hom}_S(?, T). \tag{4.5.1}$$

By definition, $\gamma(T) = S^\star$ so, using Fitting's lemma, γ takes indecomposable tilting modules over S to indecomposable projectives over $S^\star$. Moreover, the functor γ is exact on short exact sequences of modules with Δ-filtrations, so if we define $\Delta'(\lambda) = \gamma(\Delta(\lambda))$, we see that $P'(\lambda) := \gamma(T(\lambda))$ has a filtration with sections $\Delta'(\lambda)$. Then, as in [Do$_7$, A.4.7], we deduce that $S^\star$ is a quasi-hereditary algebra with indecomposable projectives $\{P'(\lambda)\}$ and standard modules $\{\Delta'(\lambda)\}$.

We will need the following known result. Although a proof of this can be deduced from the literature (using [HR, Theorem 2.1] or [CPS$_1$, Theorem 2.1(d)] together with [Do$_2$] to verify that our tilting modules coincide with the original notion of tilting module in [HR, CPS$_1$]), we include a short direct proof working purely in the framework of quasi-hereditary algebras. The argument was explained to us by S. Donkin, and we are grateful for his permission to include it here.

4.5a. **Lemma.** *Regarded as a left $S^\star$-module, T is a full tilting module for $S^\star$. Moreover, the Ringel dual $\mathrm{End}_{S^\star}(T)^{\mathrm{op}}$ of $S^\star$ relative to T is isomorphic to S.*

Proof. Let γ be the contravariant functor defined in (4.5.1) and $\Delta'(\lambda) = \gamma(\Delta(\lambda))$ for each $\lambda \in \Lambda^+$. For $M, N \in \mathrm{mod}(S)$ with Δ-filtrations, we have that $\mathrm{Ext}^i_S(M, N) \cong \mathrm{Ext}^i_{S^\star}(\gamma(N), \gamma(M))$ by [Do$_7$, A.4.8] (or rather its analogue for the functor γ). In particular, $\mathrm{Ext}^1_{S^\star}(\Delta'(\lambda), \gamma(S)) = \mathrm{Ext}^1_S(S, \Delta(\lambda)) = 0$, so that $\gamma(S)$ has a ∇'-filtration by the cohomological criterion [Do$_7$, A2.2(iii)]. Also S has a Δ-filtration so, using the fact that γ is exact on short exact sequences of modules with Δ-filtrations, we see that $\gamma(S)$ also has a Δ'-filtration. So $\gamma(S)$ is a tilting module, and more generally, passing to indecomposable summands of S, γ takes each indecomposable projective $P(\lambda)$ to a tilting module. Since $P(\lambda)$ has a Δ-filtration with $\Delta(\lambda)$ appearing with multiplicity one and all other factors being of the form $\Delta(\mu)$ for $\mu > \lambda$, $\gamma(P(\lambda))$ also has a Δ'-filtration with $\Delta'(\lambda)$ being the most dominant appearing (for the opposite ordering). So, the tilting module $\gamma(P(\lambda))$ definitely contains $T'(\lambda)$ as a summand. We deduce finally that $\gamma(S)$ is a full tilting module for $S^\star$, and from the definition of γ, $\gamma(S) = T$. This shows that T is a full tilting module for $S^\star$. Finally,

$$\mathrm{End}_{S^\star}(T)^{\mathrm{op}} = \mathrm{End}_{S^\star}(\gamma(S))^{\mathrm{op}} \cong \mathrm{End}_S(S) = S$$

using [Do7, A.4.8] once more (recall that γ is a *contravariant* functor). □

Now for the remainder of the section, we choose $\sigma \in \bar{\mathbb{F}}_q^\times$ of degree d over $\mathbb{F}_q$ and $n = kd$ for some $k \geq 1$. Let $C_{\sigma,k}$ denote the cuspidal algebra of (3.2.1). Thanks to Theorem 4.3b, the hypotheses (A1) and (A2) of §3.2 are satisfied, so $C_{\sigma,k}$ is Morita equivalent to the q^d-Schur algebra $S_{h,k} = S_{F,q^d}(h,k)$, for some fixed $h \geq k$.

Applying Ringel's theorem first to the q^d-Schur algebra $S_{h,k} = S_{F,q^d}(h,k)$, for any $h, k \geq 1$, we obtain the indecomposable tilting modules of $S_{h,k}$. We denote these modules by $\{T_h(\lambda) \,|\, \lambda \in \Lambda^+(h,k)\}$. Applying Ringel's theorem to $C_{\sigma,k}$ instead, we obtain indecomposable tilting modules $\{T(\sigma,\lambda) \mid \lambda \vdash k\}$ for $C_{\sigma,k}$. It follows immediately from (3.5a) and the definitions that

$$T(\sigma,\lambda) \cong \beta_{\sigma,h,k}(T_h(\lambda')).$$

Now, in view of Lemma 3.5d, the following statement follows from the elementary weight argument of [Do7, §3.3(1)], on applying the functor $\beta_{\sigma,h,k}$:

(4.5b) *The indecomposable tilting modules for $C_{\sigma,k}$ are precisely the indecomposable summands of $\dot{\Lambda}^\nu(\sigma)$ for all $\nu \vDash k$. Furthermore, for $\lambda \vdash k$, the module $T(\sigma,\lambda)$ occurs exactly once as a summand of $\dot{\Lambda}^\lambda(\sigma)$, and if $T(\sigma,\mu)$ is a summand of $\dot{\Lambda}^\lambda(\sigma)$ for some $\mu \vdash k$ then $\mu \geq \lambda$.*

As a corollary of (4.5b), we obtain the following alternative description of $T(\sigma,\lambda)$:

4.5c. **Lemma.** *For $\lambda \vdash k$, the module $T(\sigma,\lambda)$ can be characterized as the unique indecomposable summand of $\dot{\Lambda}^\lambda(\sigma)$ containing a submodule isomorphic to $\Delta(\sigma,\lambda)$.*

Proof. By Theorem 3.5e(i), $\dot{\Lambda}^\lambda(\sigma)$ has a unique submodule isomorphic to $\Delta(\sigma,\lambda)$. By (4.5b), $\dot{\Lambda}^\lambda(\sigma)$ has a unique summand isomorphic to $T(\sigma,\lambda)$ and for any other summand M of $\dot{\Lambda}^\lambda(\sigma)$, $\mathrm{Hom}_{C_{\sigma,k}}(\Delta(\sigma,\lambda), M) = 0$. These two statements imply the lemma. □

Next, we explain the role of Theorem 3.4a in the theory. Recall that $h \geq k$.

4.5d. **Theorem.** *The $C_{\sigma,k}$-module $T = \bigoplus_{\nu \in \Lambda(h,k)} \dot{\Lambda}^\nu(\sigma)$ is a full tilting module. Moreover, the Ringel dual $C_{\sigma,k}^\star$ of $C_{\sigma,k}$ relative to T is precisely the algebra $S_{h,k}^{\mathrm{op}}$, where $S_{h,k}$ acts on T as in Theorem 3.4a.*

Proof. (4.5b) shows immediately that T is a full tilting module. The second statement is a restatement of Theorem 3.4a. □

As consequences of Theorem 4.5d, we have the following double centralizer properties; of these, (ii) is a generalization of Takeuchi's theorem [T].

4.5e. **Theorem.** *The following double centralizer properties hold, for $h \geq k$:*

(i) $\mathrm{End}_{C_{\sigma,k}}\left(\bigoplus_{\nu\in\Lambda(h,k)} \dot{\Lambda}^{\nu}(\sigma)\right) \cong S_{h,k}$ *and* $\mathrm{End}_{S_{h,k}}\left(\bigoplus_{\nu\in\Lambda(h,k)} \dot{\Lambda}^{\nu}(\sigma)\right) \cong C_{\sigma,k}$, *where the right $S_{h,k}$-action is as in Theorem 3.4a;*

(ii) $\mathrm{End}_{C_{\sigma,k}}(M^k(\sigma)) \cong H_k$ *and* $\mathrm{End}_{H_k}(M^k(\sigma)) \cong C_{\sigma,k}$, *where* $H_k = H_{F,q^d}(\Sigma_k)$ *and the right H_k action is as in (2.5b).*

Proof. (i) Combine Theorem 4.5d with Lemma 4.5a.

(ii) Let $e \in S_{h,k}$ be the idempotent $\phi^1_{(1^k),(1^k)}$. We note that as in [Do$_7$, §3.3(1)], the indecomposable tilting module $T_h(\lambda)$ is a summand of $\Lambda^{\lambda'}(V_h)$, for any $\lambda \in \Lambda^+(h,k)$. Consequently, by (1.3b)(ii), $T_h(\lambda)$ is both a submodule and a quotient of the $S_{h,k}$-module $S_{h,k}e$. Moreover, $S_{h,k}e$ is isomorphic to $V_h^{\otimes k}$, so is (contravariantly) self-dual. These observations imply that every composition factor of both the socle and the head of $T_h(\lambda)$ belong to the head of the projective $S_{h,k}$-module $S_{h,k}e$.

Now let $T = \bigoplus_{\nu\in\Lambda(h,k)} \dot{\Lambda}^{\nu}(\sigma)$. Writing $\tilde{T}$ for the left $S_{h,k}$-module obtained from the right module T by twisting with τ, $\tilde{T}$ is a full tilting module for $S_{h,k}$ by Lemma 4.5a and Theorem 4.5d. The previous paragraph therefore shows that every composition factor of the socle and the head of $\tilde{T}$ belong to the head of $S_{h,k}e$. In other words, $V = W = \tilde{T}$ satisfy the conditions of Corollary 3.1c, taking $C = S_{h,k}, P = S_{h,k}e$ and $H = eS_{h,k}e$. We deduce at once from Corollary 3.1c that $\mathrm{End}_{S_{h,k}}(\tilde{T}) \cong \mathrm{End}_{eS_{h,k}e}(e\tilde{T})$. Switching to right actions, and using (i), we have now shown that

$$C_{\sigma,k} \cong \mathrm{End}_{eS_{h,k}e}(Te). \tag{4.5.2}$$

Now we can prove the theorem. We know already that $H_k = \mathrm{End}_{C_{\sigma,k}}(M^k(\sigma))$. As a left $C_{\sigma,k}$-module, $M^k(\sigma) \cong Te$. Moreover, by the way the action of $S_{h,k}$ on T was defined in (3.4b), the $(C_{\sigma,k}, eS_{h,k}e)$-bimodule Te is isomorphic to the $(C_{\sigma,k}, H_k)$-bimodule $M^k(\sigma)$, if we identify H_k with $eS_{h,k}e$ so that $T_w \mapsto \kappa(T_w^{\#})$ for each $w \in \Sigma_k$, where κ is as in (1.2b). In view of this,

$$\mathrm{End}_{H_k}(M^k(\sigma)) \cong \mathrm{End}_{eS_{h,k}e}(Te)$$

which is isomorphic to $C_{\sigma,k}$ thanks to (4.5.2). □

We end with the non-defining characteristic analogues of [Do$_5$, Lemma 3.4(i)] and [MP, Corollary 2.3]. For $\lambda \vdash k$, let $P(\sigma,\lambda)$ denote the projective cover of $L(\sigma,\lambda)$ in the category $\mathrm{mod}(C_{\sigma,k})$. Similarly, for $\mu \in \Lambda^+(h,k)$, let $P_h(\mu)$ denote the projective cover of $L_h(\mu)$ in the category $\mathrm{mod}(S_{h,k})$. Obviously we have that:

$$P(\sigma,\lambda) \cong \beta_{\sigma,h,k}(P_h(\lambda')). \tag{4.5.3}$$

4.5f. **Theorem.** *For $\nu \in \Lambda(h,k)$,*

$$\dot{Z}^\nu(\sigma) \cong \bigoplus_{\lambda \vdash k} P(\sigma, \lambda')^{\oplus m_{\lambda,\nu}}, \tag{4.5.4}$$

$$\dot{\Lambda}^\nu(\sigma) \cong \bigoplus_{\lambda \vdash k} T(\sigma, \lambda)^{\oplus m_{\lambda,\nu}}, \tag{4.5.5}$$

where $m_{\lambda,\nu}$ is equal to the dimension of the ν-weight space of $L_h(\lambda)$.

Proof. Using Fitting's lemma, there is a unique decomposition $1 = \sum_{\lambda \in \Lambda^+(h,k)} e_\lambda$ of the identity of $S_{h,k}$ into orthogonal idempotents such that for each λ,

$$S_{h,k}e_\lambda \cong P_h(\lambda)^{\oplus \dim L_h(\lambda)}.$$

We observe that $\dim L_h(\lambda) = \dim \mathrm{Hom}_{S_{h,k}}(S_{h,k}e_\lambda, L_h(\lambda)) = \dim e_\lambda L_h(\lambda)$. This shows that $e_\lambda L_h(\lambda) = L_h(\lambda)$. Now, for any $\nu \in \Lambda(h,k)$, $\phi^1_{\nu,\nu} \in S_{h,k}$ is an idempotent which obviously commutes with each e_λ. So, $\phi^1_{\nu,\nu}e_\lambda$ is an idempotent. Observe that

$$\mathrm{Hom}_{S_{h,k}}(S_{h,k}\phi^1_{\nu,\nu}e_\lambda, L_h(\lambda)) \cong \phi^1_{\nu,\nu}e_\lambda L_h(\lambda) = \phi^1_{\nu,\nu}L_h(\lambda),$$

which is the ν-weight space of $L_h(\lambda)$. So, $S_{h,k}\phi^1_{\nu,\nu}e_\lambda$ is a direct sum of precisely $m_{\lambda,\nu}$ copies of $P_h(\lambda)$.

To prove (4.5.4), we have so far shown that $S_{h,k}\phi^1_{\nu,\nu} \cong \bigoplus_{\lambda \vdash k} P_h(\lambda)^{m_{\lambda,\nu}}$. Now apply the equivalence of categories $\beta_{\sigma,h,k}$, using (4.5.3) and the fact observed in the proof of Lemma 3.5d that $\beta_{\sigma,h,k}(S_{h,k}\phi^1_{\nu,\nu}) \cong \dot{Z}^\nu(\sigma)$.

Now consider (4.5.5). Certainly by (4.5b), all indecomposable summands of $T = \bigoplus_{\nu \in \Lambda(h,k)} \dot{\Lambda}^\nu(\sigma)$ are of the form $T(\sigma,\lambda)$, $\lambda \vdash k$. Combining this with Theorem 3.4a and Fitting's lemma, we see that there is some permutation π of the set of partitions of k such that, for each $\lambda \vdash k$, Te_λ is the largest summand of T isomorphic to a direct sum of copies of $T(\sigma, \pi(\lambda))$. Recall from the definition of the action of $\phi^1_{\nu,\nu}$ from Theorem 3.4a that $T\phi^1_{\nu,\nu} = \dot{\Lambda}^\nu(\sigma)$. So, $T\phi^1_{\nu,\nu}e_\lambda$ is the largest summand of $\dot{\Lambda}^\nu(\sigma)$ isomorphic to a direct sum of copies of $T(\sigma,\pi(\lambda))$. Moreover, using the first paragraph, we know that in fact $T\phi^1_{\nu,\nu}e_\lambda$ is a direct sum of precisely $m_{\lambda,\nu}$ copies of $T(\sigma,\pi(\lambda))$. In other words, we know that for each $\nu \vDash k$:

$$\dot{\Lambda}^\nu(\sigma) \cong \bigoplus_{\lambda \vdash k} T(\sigma, \pi(\lambda))^{\oplus m_{\lambda,\nu}}. \tag{4.5.6}$$

It therefore remains to prove that $\pi(\lambda) = \lambda$ for each $\lambda \vdash k$. We prove this by downward induction on the dominance order on λ. The induction starts with $\lambda = (k)$; here, $\pi((k)) = (k)$ immediately from (4.5.6), since $\Lambda^k(\sigma) = T(\sigma,(k))$. Now take any $\mu < (k)$ and suppose we have proved inductively that π fixes all more dominant partitions. Using the inductive hypothesis, (4.5.6) tells us that

$$\dot{\Lambda}^\mu(\sigma) \cong T(\sigma, \pi(\mu)) \oplus \bigoplus_{\lambda > \mu} T(\sigma, \lambda)^{\oplus m_{\lambda,\mu}}.$$

Finally, we know by (4.5b) that $T(\sigma,\mu)$ appears as a summand of $\dot{\Lambda}^\mu(\sigma)$, so we must have that $T(\sigma,\mu) \cong T(\sigma,\pi(\mu))$, whence $\pi(\mu) = \mu$ as required. □

Chapter 5

The affine general linear group

In this chapter we prove results that can be regarded as the modular analogues of the branching rules of Zelevinsky [Z, Theorem 13.5] and Thoma [Th]. Following the idea of Zelevinsky, we study the affine general linear group $AGL_n(\mathbb{F}_q)$ with the same methods as we have developed so far for $GL_n(\mathbb{F}_q)$. Roughly speaking, our main result relates restriction from $GL_n(\mathbb{F}_q)$ to $AGL_{n-1}(\mathbb{F}_q)$ to restriction from quantum GL_n to quantum GL_{n-1}. As an application, we obtain a new dimension formula for irreducible modular representations of $GL_n(\mathbb{F}_q)$, in terms of weight space dimensions of irreducible modules over quantum GL_n.

5.1. Levels and the branching rule from AGL_n to GL_n

For a group G, $\mathcal{I}_G$ will denote the trivial FG-module, where F is as usual our fixed algebraically closed field of characteristic p coprime to q. Also, for a subgroup $H \subset G$ and an FH-module M (resp. an FG-module N), we will often write $M \uparrow^G$ (resp. $N \downarrow_H$) for $\mathrm{ind}_H^G M$ (resp. $\mathrm{res}_H^G N$).

Let W_n denote an n-dimensional vector space over $\mathbb{F}_q$ with basis $w_1, \dots, w_n$. Let G_n denote $GL_n(\mathbb{F}_q)$ acting naturally on W_n. Let H_n denote the *affine general linear group* $AGL_n(\mathbb{F}_q)$, which is the semi-direct product $G_n W_n$ of W_n (regarded now just as an Abelian group) by G_n. By convention, we allow the notations G_0, H_0 and W_0, all of which denote groups with one element.

We make some remarks about the representations of the Abelian group W_n over F. Let X_n denote the set of irreducible characters $W_n \to F^\times$, regarded as an n-dimensional $\mathbb{F}_q$-vector space via $(\varepsilon + \chi)(w) = \varepsilon(w)\chi(w)$ and $(c\varepsilon)(w) = \varepsilon(cw)$ for $\varepsilon, \chi \in X_n, c \in \mathbb{F}_q, w \in W_n$. To describe a parameterization of these characters, let $\chi : \mathbb{F}_q \to F^\times$ be the non-trivial character fixed after (2.5.4). For any element θ of the $\mathbb{F}_q$-linear dual W_n^*, let ε_θ denote the character sending $w \in W_n$ to $\chi(\theta(w)) \in F^\times$. Then, X_n is precisely the set $\{\varepsilon_\theta \mid \theta \in W_n^*\}$. We let $\theta_1, \dots, \theta_n$ denote the basis of W_n^* dual to $w_1, \dots, w_n$, and set $\varepsilon_i = \varepsilon_{\theta_i}$ to obtain a basis $\varepsilon_1, \dots, \varepsilon_n$ for the $\mathbb{F}_q$-vector space X_n.

The group G_n acts on the characters X_n by $(g\varepsilon)(w) = \varepsilon(g^{-1}w)$, for $g \in G_n, w \in$

$W_n, \varepsilon \in X_n$. Under this action, the trivial character $0 \in X_n$ is fixed by G_n and all other characters are permuted transitively. An elementary calculation shows that the group $C_{G_n}(\varepsilon_n)$ is isomorphic to H_{n-1}, embedded into G_n as the subgroup of matrices of the form

$$\left[\begin{array}{ccc|c} & * & & * \\ \hline 0 & \dots & 0 & 1 \end{array}\right].$$

We always identify H_{n-1} with this subgroup of G_n. Thus, we have a chain of subgroups

$$1 = H_0 \subset G_1 \subset H_1 \subset G_2 \subset H_2 \subset \dots.$$

Given an FH_n-module M and $\varepsilon \in X_n$, let

$$M_\varepsilon = \{m \in M \mid wm = \varepsilon(w)m \text{ for all } w \in W_n\}$$

be the corresponding *weight space*. Since FW_n is a semisimple algebra, M decomposes as a direct sum of such weight spaces. Observe that the action of G_n on M induces an action on the weight spaces, so that $gM_\varepsilon = M_{g\varepsilon}$ for all $g \in G_n, \varepsilon \in X_n$. In particular, the 0-weight space M_0 is stable under the action of G_n, while the ε_n-weight space M_{ε_n} of M is stable under the action of the subgroup $H_{n-1} < G_n$, since H_{n-1} centralizes ε_n.

Now we introduce some functors. First, for $n \geq 0$, we have functors

$$f_0^n : \text{mod}(FH_n) \to \text{mod}(FG_n),$$
$$e_0^n : \text{mod}(FG_n) \to \text{mod}(FH_n).$$

For these, f_0^n is defined on an object M by $M \mapsto M_0$, the zero weight space of M, which we observed in the previous paragraph is G_n-stable. On a morphism, f_0^n is defined simply to be its restriction to zero weight spaces. The functor e_0^n is defined on an object M to be the same vector space, but regarded as an FH_n-module by extending the action of G_n on M to H_n by letting W_n act trivially, and on a morphism by simply regarding the morphism as a homomorphism over H_n instead of G_n. Next, for $n \geq 1$, we have functors

$$f_+^n : \text{mod}(FH_n) \to \text{mod}(FH_{n-1}),$$
$$e_+^n : \text{mod}(FH_{n-1}) \to \text{mod}(FH_n).$$

On an object M, f_+^n sends M to its ε_n-weight space M_{ε_n}, which is stable under the action of H_{n-1}; on a morphism, f_+^n is defined by restriction. The functor e_+^n is defined to be the composite of the inflation functor from H_{n-1} to $H_{n-1}W_n$, with the action of W_n being via the character ε_n, followed by ordinary induction from $H_{n-1}W_n$ to H_n. Finally, for $n \geq 1$ and $1 \leq i \leq n$, we have functors

$$f_i^n : \text{mod}(FH_n) \to \text{mod}(FG_{n-i}),$$
$$e_i^n : \text{mod}(FG_{n-i}) \to \text{mod}(FH_n).$$

These are defined inductively by $f_i^n = f_{i-1}^{n-1} \circ f_+^n$ and $e_i^n = e_+^n \circ e_{i-1}^{n-1}$. By convention, if $i > n$, the functor f_i^n denotes the zero functor. It will usually be obvious from context which group H_n we have in mind, so we will from now on drop the index n from our notation for the functors e_+^n, e_i^n, f_i^n and f_+^n.

5.1a. **Lemma.** (i) *For any $n \geq 1$, the functors*

$$f_0 \oplus f_+ : \mathrm{mod}(FH_n) \to \mathrm{mod}(FG_n) \times \mathrm{mod}(FH_{n-1}),$$
$$e_0 \oplus e_+ : \mathrm{mod}(FG_n) \times \mathrm{mod}(FH_{n-1}) \to \mathrm{mod}(FH_n)$$

are mutually inverse equivalences of categories. Here, $f_0 \oplus f_+$ denotes the functor $? \mapsto (f_0\, ?, f_+\, ?)$ and $e_0 \oplus e_+$ is the functor $(?, ?') \mapsto (e_0\, ?) \oplus (e_+\, ?')$.

(ii) *For any $n \geq 0$, the functors*

$$f_0 \oplus f_1 \oplus \cdots \oplus f_n : \mathrm{mod}(FH_n) \to \mathrm{mod}(FG_n) \times \mathrm{mod}(FG_{n-1}) \times \cdots \times \mathrm{mod}(FG_0),$$
$$e_0 \oplus e_1 \oplus \cdots \oplus e_n : \mathrm{mod}(FG_n) \times \mathrm{mod}(FG_{n-1}) \times \cdots \times \mathrm{mod}(FG_0) \to \mathrm{mod}(FH_n)$$

are mutually inverse equivalences of categories.

Proof. (i) As in [Z, §13.1], this is a special case of a general result about representations of a semi-direct product GW where W is an Abelian normal subgroup of GW, see [Se, 8.2]. Although the argument in *loc. cit.* is in characteristic 0, it applies equally well to our case since FW is a semisimple algebra.

(ii) Apply (i) and induction on n. □

With Lemma 5.1a as our motivation, we now make the basic definition for understanding the representation theory of H_n. Say that an FH_n-module M belongs to the *ith level* if $f_j M = 0$ for all $j \neq i$. We will also refer to the FH_n-module $e_i \circ f_i(M)$ as the *ith level of M*. By the lemma, any FG_n-module M splits uniquely as the direct sum of its levels.

We will now write $\mathcal{C}_{ss,p'}^n$ for the set $\mathcal{C}_{ss,p'}$ of representatives of the p-regular classes of semisimple elements G_n chosen in §2.1. We allow the notation $\mathcal{C}_{ss,p'}^0$, which is the set containing just one element, namely, the identity element $1 \in G_0$; the composition $\kappa(1)$ is then just the zero composition (0).

Applying Lemma 5.1a(ii), the irreducible FH_n-modules belonging to the ith level are precisely the modules $e_i L$ as L runs over the irreducible FG_{n-i}-modules. We obtain immediately from (4.4b) the following description of the irreducible FH_n-modules:

(5.1b) *The set $\left\{ e_i L(s, \underline{\lambda}) \mid 0 \leq i \leq n, s \in \mathcal{C}_{ss,p'}^{n-i}, \underline{\lambda} \vdash \kappa(s) \right\}$ is a complete set of non-isomorphic irreducible FH_n-modules.*

Our aim in the remainder of the chapter is to understand induction and restriction of irreducibles between GL and AGL. To start off with, we have the basic:

5.1c. **Lemma.** *The following pairs of functors are isomorphic:*
(i) $\mathrm{res}^{G_n}_{H_{n-1}}$ *and* $f_+ \circ \mathrm{ind}^{H_n}_{G_n} : \mathrm{mod}(FG_n) \to \mathrm{mod}(FH_{n-1})$;
(ii) $\mathrm{ind}^{H_n}_{G_n}$ *and* $e_0 \oplus (e_+ \circ \mathrm{res}^{G_n}_{H_{n-1}}) : \mathrm{mod}(FG_n) \to \mathrm{mod}(FH_n)$;
(iii) $\mathrm{ind}^{G_n}_{H_{n-1}}$ *and* $\mathrm{res}^{H_n}_{G_n} \circ e_+ : \mathrm{mod}(FH_{n-1}) \to \mathrm{mod}(FG_n)$;
(iv) $\mathrm{res}^{H_n}_{G_n}$ *and* $f_0 \oplus (\mathrm{ind}^{G_n}_{H_{n-1}} \circ f_+) : \mathrm{mod}(FH_n) \to \mathrm{mod}(FG_n)$.

Proof. We only need to prove (i) and (ii); then, (iii) and (iv) follow immediately on taking adjoints. The subgroup $W_n \subset H_n$ is a set of H_n/G_n-coset representatives. Since FW_n is a semisimple algebra, we can pick a basis $\{z_\varepsilon \mid \varepsilon \in X_n\}$ for FW_n such that $wz_\varepsilon = \varepsilon(w)z_\varepsilon$ for each $w \in W_n$ and $\varepsilon \in X_n$. Then, for any FG_n-module M, we see that $\mathrm{ind}^{H_n}_{G_n} M = FH_n \otimes_{FG_n} M$ can be written as a direct sum:

$$\mathrm{ind}^{H_n}_{G_n} M = \bigoplus_{\varepsilon \in X_n} z_\varepsilon \otimes M.$$

Then, $f_0(\mathrm{ind}^{H_n}_{G_n} M) = z_0 \otimes M \cong M$ and $f_+(\mathrm{ind}^{H_n}_{G_n} M) = z_{\varepsilon_n} \otimes M \cong M \downarrow_{H_{n-1}}$. We deduce that there are isomorphisms of functors:

$$f_0 \circ \mathrm{ind}^{H_n}_{G_n} \cong \mathrm{id}_{\mathrm{mod}(FG_n)}, \qquad f_+ \circ \mathrm{ind}^{H_n}_{G_n} \cong \mathrm{res}^{G_n}_{H_{n-1}}.$$

In particular, this proves (i), while (ii) follows on applying Lemma 5.1a(i) and the properties above. □

Before the next theorem, we introduce some further notation. Given $m, n \geq 0$, let $S_{m,n}$ and $Z_{m,n}$ be the subgroups of G_{m+n} consisting of all matrices of the form:

$$S_{m,n} : \left[\begin{array}{c|cc} *_m & \multicolumn{2}{c}{*} \\ \hline & 1 & * \\ 0 & & \ddots \\ & 0 & 1 \end{array}\right], \qquad Z_{m,n} : \left[\begin{array}{c|c} I_m & * \\ \hline 0 & I_n \end{array}\right].$$

We view the group U_n of all upper uni-triangular matrices in G_n as a subgroup of $S_{m,n}$, embedded in the obvious way into the bottom right hand corner of the matrices. Doing this, we can regard the Gelfand-Graev idempotent $\gamma_n \in FU_n$ of (2.5.5) as an element of $FS_{m,n}$. We also recall for use shortly that $F\gamma_n$ is a one dimensional FU_n-module, and the induced module $(F\gamma_n) \uparrow^{G_n}$ is the Gelfand-Graev representation Γ_n of FG_n.

Now, there is an obvious surjective homomorphism

$$S_{m,n} \to G_m \times U_n \tag{5.1.1}$$

with kernel $Z_{m,n}$. Define the functor

$$\mathrm{infl}^{S_{m,n}}_{G_m} : \mathrm{mod}(FG_m) \to \mathrm{mod}(FS_{m,n})$$

to be the composite of the functor $? \boxtimes F\gamma_n : \mathrm{mod}(FG_m) \to \mathrm{mod}(F(G_m \times U_n))$ followed by natural inflation functor along the surjection (5.1.1). Note that $S_{m,n+1}$ is a subgroup of H_{m+n}.

5.1d. **Lemma.** *For any $m, n \geq 0$, the functor $e_n : \mathrm{mod}(FG_m) \to \mathrm{mod}(FH_{m+n})$ is isomorphic to the composite functor $\mathrm{ind}_{S_{m,n+1}}^{H_{m+n}} \circ \mathrm{infl}_{G_m}^{S_{m,n+1}}$.*

Proof. We proceed by induction on n, the case $n = 0$ being trivial. Recalling that $e_n = e_+ \circ e_{n-1}$, the inductive step reduces at once to proving that there is an isomorphism of functors:

$$e_+ \circ \mathrm{ind}_{S_{m,n}}^{H_{m+n-1}} \circ \mathrm{infl}_{G_m}^{S_{m,n}} \cong \mathrm{ind}_{S_{m,n+1}}^{H_{m+n}} \circ \mathrm{infl}_{G_m}^{S_{m,n+1}}.$$

To see this, it is easier to show instead that the adjoints of each of these functors are isomorphic.

The adjoint functor to e_+ is $f_+ : \mathrm{mod}(FH_{m+n}) \to \mathrm{mod}(FH_{m+n-1})$. This can be viewed simply as left multiplication by the idempotent

$$z_{\varepsilon_{m+n}} = \frac{1}{|W_{m+n}|} \sum_{w \in W_{m+n}} \varepsilon_{m+n}(-w)w \in FW_{n+m} \subset FH_{m+n}$$

corresponding to the character $\varepsilon_{m+n} \in X_{m+n}$. The adjoint of $\mathrm{ind}_{S_{m,n}}^{H_{m+n-1}} \circ \mathrm{infl}_{G_m}^{S_{m,n}}$ is given by first restriction from H_{m+n-1} to $S_{m,n}$, followed by multiplication by the idempotent $\zeta_{m,n}\gamma_n$, where

$$\zeta_{m,n} = \frac{1}{|Z_{m,n}|} \sum_{z \in Z_{m,n}} z \in FS_{m,n}$$

and we are viewing γ_n as an element of $FS_{m,n}$ as explained above. Hence, the adjoint of the composite functor $e_+ \circ \mathrm{ind}_{S_{m,n}}^{H_{m+n-1}} \circ \mathrm{infl}_{G_m}^{S_{m,n}}$ is given on objects just by multiplication by the idempotent $\zeta_{m,n}\gamma_n z_{\varepsilon_{m+n}} \in FH_{m+n}$, and by restriction on morphisms.

On the other hand, the adjoint of the functor $\mathrm{ind}_{S_{m,n+1}}^{H_{m+n}} \circ \mathrm{infl}_{G_m}^{S_{m,n+1}}$ is given on objects by multiplication by the idempotent $\zeta_{m,n+1}\gamma_{n+1} \in FH_{m+n}$. Now an easy matrix calculation in the group algebra FH_{m+n} reveals that $\zeta_{m,n+1}\gamma_{n+1} = \zeta_{m,n}\gamma_n z_{\varepsilon_{m+n}}$. Hence our functors are isomorphic. □

Now we obtain the following fundamental result:

5.1e. **Theorem.** *For any $m, n \geq 0$, the following functors are isomorphic:*

$$R_{G_m \times G_n}^{G_{m+n}}(? \boxtimes \Gamma_n) : \mathrm{mod}(FG_m) \to \mathrm{mod}(FG_{m+n})$$

$$\mathrm{res}_{G_{m+n}}^{H_{m+n}} \circ e_n : \mathrm{mod}(FG_m) \to \mathrm{mod}(FG_{m+n}).$$

Proof. We will view all of the groups $G_{m+n}, S_{m,n}, S_{m,n+1}$ as naturally embedded subgroups of $H_{m+n} \subset G_{m+n+1}$. Also, write $P_{m,n}$ for the standard parabolic subgroup of G_{m+n} with Levi factor $G_m \times G_n$ and unipotent radical $Z_{m,n}$.

Take any FG_m-module M. We first observe that as $(F\gamma_n) \uparrow^{G_n} \cong \Gamma_n$, there is an isomorphism of functors:

$$R^{G_{m+n}}_{G_m \times G_n}(\,?\, \boxtimes \Gamma_n) \cong \mathrm{ind}^{G_{m+n}}_{S_{m,n}} \circ \mathrm{infl}^{S_{m,n}}_{G_m}. \tag{5.1.2}$$

Now, there is a factorization $H_{m+n} = G_{m+n}S_{m,n+1}$, and $G_{m+n} \cap S_{m,n+1} = S_{m,n}$. So the Mackey theorem gives an isomorphism of functors

$$\mathrm{res}^{H_{m+n}}_{G_{m+n}} \circ \mathrm{ind}^{H_{m+n}}_{S_{m,n+1}} \cong \mathrm{ind}^{G_{m+n}}_{S_{m,n}} \circ \mathrm{res}^{S_{m,n+1}}_{S_{m,n}}. \tag{5.1.3}$$

Since $(F\gamma_{n+1}) \downarrow_{U_n} \cong F\gamma_n$, there is an isomorphism of functors

$$\mathrm{res}^{S_{m,n+1}}_{S_{m,n}} \circ \mathrm{infl}^{S_{m,n+1}}_{G_m} \cong \mathrm{infl}^{S_{m,n}}_{G_m}. \tag{5.1.4}$$

Now combine (5.1.2), (5.1.3), (5.1.4) and Lemma 5.1d to complete the proof. □

Theorem 5.1e has a number of important consequences. First, we obtain a proof of Gelfand's theorem [Ge]:

5.1f. **Corollary.** *Take $\sigma \in \bar{\mathbb{F}}_q^\times$ of degree d over $\mathbb{F}_q$. Then,*
(i) $M(\sigma) \uparrow^{H_d} \cong e_0 M(\sigma) \oplus e_d \mathcal{I}_{G_0}$;
(ii) $M(\sigma) \downarrow_{H_{d-1}} \cong e_{d-1}\mathcal{I}_{G_0}$.

Proof. (i) For any $0 \leq i \leq d$ and any irreducible FG_{d-i}-module L, compute using Frobenius reciprocity and Theorem 5.1e:

$$\begin{aligned}\mathrm{Hom}_{H_d}(e_i L, M(\sigma) \uparrow^{H_d}) &\cong \mathrm{Hom}_{G_d}((e_i L) \downarrow_{G_d}, M(\sigma))\\ &\cong \mathrm{Hom}_{G_d}(R^{G_d}_{G_{d-i}\times G_i}(L \boxtimes \Gamma_i), M(\sigma))\\ &\cong \mathrm{Hom}_{G_{d-i}\times G_i}(L \boxtimes \Gamma_i, {}^*R^{G_d}_{G_{d-i}\times G_i} M(\sigma)).\end{aligned}$$

Since $M(\sigma)$ is cuspidal and irreducible, this is zero unless *either* $i = 0, L = M(\sigma)$ *or* $i = d, L = \mathcal{I}_{G_0}$. In the former case, the resulting hom space is obviously one dimensional, while in the latter case, it is one dimensional thanks to Corollary 2.5e(ii). This shows that the socle of $M(\sigma) \uparrow^{H_d}$ has just two irreducible constituents, namely, $e_0 M(\sigma)$ and $e_d \mathcal{I}_{G_0}$. Finally, by a dimension calculation using (2.4.1), we must have in fact that $M(\sigma) \uparrow^{H_d}$ is equal to its socle.

(ii) Apply (i) and Lemma 5.1c. □

The next corollary describes the restriction of an arbitrary irreducible FH_n-module to FG_n.

5.1g. **Corollary.** *If $M = e_i L$ is an irreducible FH_n-module belonging to the ith level, then*

$$M \downarrow_{G_n} \cong R^{G_n}_{G_{n-i}\times G_i}(L \boxtimes \Gamma_i).$$

Proof. This is immediate from Theorem 5.1e. □

5.1h. **Remark.** Using Corollary 5.1g in characteristic 0, one can obtain alternative proofs of the branching rules of Zelevinsky [Z, Theorem 13.5] and Thoma [Th]. We sketch the argument. First, using (2.3f) one can decompose $R^{G_n}_{G_{n-i}\times G_i}(L \boxtimes L')$ for any irreducible modules L, L': in characteristic zero this reduces to the Littlewood-Richardson rule. Second, one knows how to decompose Γ_i into a direct sum of irreducibles in characteristic 0: as in Theorem 2.5d, one gets all irreducible modules $L(s, \lambda_s)$ for all $s \in \mathcal{C}_{ss}$, where for s of the form (2.1.1) $\lambda_s = ((1^{k_1}), \dots, (1^{k_a}))$, each appearing with multiplicity one. Hence, using Corollary 5.1g, one can decompose the restriction of any irreducible KH_n-module to KG_n. Finally, one uses Frobenius reciprocity and Lemma 5.1c(i) to deduce the formula for the restriction of any irreducible KG_n-module to KH_{n-1}. This gives the branching rules of Zelevinsky, and some further combinatorial argument deduces Thoma's branching rule from this.

5.2. Affine induction operators

As motivation, we recall the definition of Green's induction operator $\diamond$, originally introduced in [G₁]. Take integers $m, n \geq 0$. Given an FG_m-module M and an FG_n-module N, let $M \diamond N$ denote the FG_{m+n}-module

$$M \diamond N = R^{G_{n+m}}_{G_n \times G_m}(M \boxtimes N).$$

The resulting operator $\diamond$ allows one to 'multiply' two GL-modules to obtain a new GL-module. We will define two new operators $\diamond_\ell$ and $\diamond_r$, which will allow one to 'multiply' an AGL-module by a GL-module on the left and on the right, respectively, to obtain a new AGL-module.

To define $\diamond_\ell$, let M be an FG_m-module and N be an FH_n-module. Consider the subgroups $Q_{m,n}$ and $X_{m,n}$ of $H_{m+n} \subset G_{m+n+1}$ consisting of all matrices of the form

$$Q_{m,n} : \left[\begin{array}{ccc|ccc|c} & *_m & & & * & & * \\ \hline & 0 & & & *_n & & * \\ \hline 0 & \dots & 0 & 0 & \dots & 0 & 1 \end{array}\right] \qquad X_{m,n} : \left[\begin{array}{ccc|ccc|c} & I_m & & & * & & * \\ \hline & 0 & & & I_n & & 0 \\ \hline 0 & \dots & 0 & 0 & \dots & 0 & 1 \end{array}\right]$$

There is an obvious surjective homomorphism

$$Q_{m,n} \to G_m \times H_n$$

with kernel $X_{m,n}$. Now define $M \diamond_\ell N$ to be the FH_{m+n}-module obtained by first inflating the $F(G_m \times H_n)$-module $M \boxtimes N$ to $Q_{m,n}$ along this surjection, then inducing as usual from $Q_{m,n}$ to H_{m+n}.

Next, to define $\diamond_r$, let M be an FH_m-module and N be an FG_n-module. Consider the subgroups $R_{m,n}$ and $Y_{m,n}$ of $H_{m+n} \subset G_{m+n+1}$ consisting of all matrices of the form

$$R_{m,n}: \left[\begin{array}{ccc|ccc|c} & *_m & & & * & & * \\ \hline & 0 & & & *_n & & 0 \\ \hline 0 & \dots & 0 & 0 & \dots & 0 & 1 \end{array}\right] \qquad Y_{m,n}: \left[\begin{array}{ccc|ccc|c} & I_m & & & * & & 0 \\ \hline & 0 & & & I_n & & 0 \\ \hline 0 & \dots & 0 & 0 & \dots & 0 & 1 \end{array}\right]$$

This time, there is a surjective homomorphism

$$R_{m,n} \to H_m \times G_n$$

with kernel $Y_{m,n}$. Define $M \diamond_r N$ to be the FH_{m+n}-module obtained by first inflating the $F(H_m \times G_n)$-module $M \boxtimes N$ to $R_{m,n}$ along this surjection, then inducing as usual from $R_{m,n}$ to H_{m+n}.

The next lemma explains our interest in the operators $\diamond_\ell$ and $\diamond_r$.

5.2a. **Lemma.** *Given an FG_m-module M and an FG_n-module N,*

$$(M \diamond N) \downarrow_{H_{m+n-1}} \cong M \diamond_\ell (N \downarrow_{H_{n-1}}) \oplus (M \downarrow_{H_{m-1}}) \diamond_r N.$$

Proof. Let $P_{m,n}$ denote the standard parabolic subgroup of G_{m+n} with Levi factor $G_m \times G_n$, which is the stabilizer of the subspace of W_{m+n} spanned by $\{w_1, \dots, w_m\}$. The subgroup H_{m+n-1} has precisely two orbits on the set of m-dimensional subspaces of W_{m+n}, with representatives the subspaces spanned by $\{w_1, \dots, w_{m-1}, w_m\}$ and $\{w_1, \dots, w_{m-1}, w_{m+n}\}$. So there are two $H_{m+n-1}\backslash G_{m+n}/P_{m,n}$-double cosets in G_{m+n}, with representatives 1 and π, where π is the permutation matrix corresponding to the cycle $(m+n\ \ m+n-1\ \ \dots\ \ m)$ (so $\pi w_m = w_{m+n}$). Now the Mackey formula gives us that $(M \diamond N) \downarrow_{H_{m+n-1}}$ is isomorphic to

$$(M\#N) \downarrow_{P_{m,n} \cap H_{m+n-1}} \uparrow^{H_{m+n-1}} \oplus (\mathrm{conj}_\pi(M\#N)) \downarrow_{{}^\pi P_{m,n} \cap H_{m+n-1}} \uparrow^{H_{m+n-1}},$$

where $M\#N$ denotes the $FP_{m,n}$-module obtained by inflation from the $F(G_m \times G_n)$-module $M \boxtimes N$. Observe that $P_{m,n} \cap H_{m+n-1} = Q_{m,n-1}$, so that the first term gives us precisely $M \diamond_\ell (N \downarrow_{H_{n-1}})$, and ${}^\pi P_{m,n} \cap H_{m+n-1} = R_{m-1,n}$ so the second term is isomorphic to $(M \downarrow_{H_{m-1}}) \diamond_r N$. □

5.2b. **Theorem.** (i) *Given an FG_m-module M and an FH_n-module N,*

$$f_0(M \diamond_\ell N) \cong M \diamond (f_0 N),$$
$$f_+(M \diamond_\ell N) \cong M \diamond_\ell (f_+ N).$$

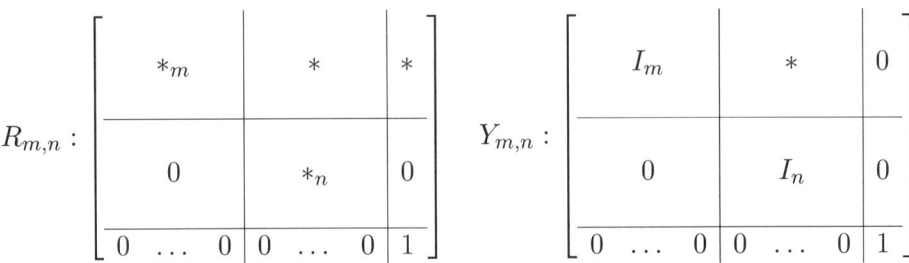

(ii) *Given an FH_m-module M and an FG_n-module N,*

$$f_0(M \diamond_r N) \cong (f_0 M) \diamond N,$$
$$f_+(M \diamond_r N) \cong (f_0 M) \diamond_\ell (N \downarrow_{H_{n-1}}) \oplus (f_+ M) \diamond_r N.$$

Proof. We just explain how to prove the final formula

$$f_+(M \diamond_r N) \cong (f_0 M) \diamond_\ell (N \downarrow_{H_{n-1}}) \oplus (f_+ M) \diamond_r N \tag{5.2.1}$$

appearing in (ii). The other three formulae are proved in similar (actually much easier) ways and we omit the details.

We shall identify W_m (resp. W_n) with the subspace of W_{m+n} spanned by $w_1, \dots, w_m$ (resp. $w_{m+1}, \dots, w_{m+n}$), so $W_{m+n} = W_m \oplus W_n$. Identify X_{m+n} with $X_m \oplus X_n$ in a similar way. We also need some notation for coset representatives, so write $P_{m,n}$ for the standard parabolic subgroup of G_{m+n} with Levi factor $G_m \times G_n$. Let $\Omega_{m,n}$ denote the set of $(m+n)$-tuples:

$$\Omega_{m,n} = \left\{ \underline{i} = (i_1, i_2, \dots, i_{m+n}) \;\middle|\; \begin{array}{l} \{i_1, i_2, \dots, i_{m+n}\} = \{1, 2, \dots, m+n\}, \\ i_k < i_{k+1} \text{ unless } k = m \end{array} \right\}.$$

For $\underline{i} = (i_1, \dots, i_{m+n}) \in \Omega_{m,n}$, let $w_{\underline{i}} \in \Sigma_{m+n} \subset G_{m+n}$ be the permutation matrix corresponding to the permutation $j \mapsto i_j$ for $1 \leq j \leq m+n$. We note that $\{w_{\underline{i}} \mid \underline{i} \in \Omega_{m,n}\}$ is precisely the set of distinguished $\Sigma_{m+n}/\Sigma_m \times \Sigma_n$-coset representatives. Also define

$$U_{\underline{i}} = \{u \in U_{m+n} \mid u_{i_j i_k} = 0 \text{ for every } 1 \leq j < k \leq m+n \text{ with } i_j < i_k\}.$$

Finally, let π be the permutation matrix for the cycle $(m+n \; m+n-1 \; \dots \; m+1 \; m)$ of Σ_{m+n}. Now, regarding all of H_{m+n-1}, $Q_{m,n-1}$ and $R_{m-1,n}$ as naturally embedded subgroups of $G_{m+n} \subset H_{m+n}$, we claim:

(5.2c) (i) *$\{u w_{\underline{i}} y \mid \underline{i} \in \Omega_{m,n}, u \in U_{\underline{i}}, y \in W_n\}$ is a set of $H_{m+n}/R_{m,n}$-coset representatives.*

(ii) *$\{u w_{\underline{i}} \mid \underline{i} \in \Omega_{m,n}, i_{m+n} = m+n, u \in U_{\underline{i}}\}$ is a set of $H_{m+n-1}/Q_{m,n-1}$-coset representatives.*

(iii) *$\{u w_{\underline{i}} \pi^{-1} \mid \underline{i} \in \Omega_{m,n}, i_m = m+n, u \in U_{\underline{i}}\}$ is a set of $H_{m+n-1}/R_{m-1,n}$-coset representatives.*

To prove (5.2c), we first observe by the Bruhat decomposition that $\{u w_{\underline{i}} \mid \underline{i} \in \Omega_{m,n}, u \in U_{\underline{i}}\}$ is a set of $G_{m+n}/P_{m,n}$-coset representatives. Parts (i) and (ii) follow easily from this. To see (iii), let W'_n denote the set

$$\{w \in U_{m+n} \mid \text{for } 1 \leq j < k \leq m+n, \; w_{jk} = 0 \text{ unless } j \geq m, k = m+n\}.$$

As a variant of (i), the set $\{u w_{\underline{i}} y \mid \underline{i} \in \Omega_{m-1,n+1}, i_{m+n} = m+n, u \in U_{\underline{i}}, y \in W'_n\}$ is a set of $H_{m+n-1}/R_{m-1,n}$-coset representatives, working inside the group G_{m+n}. An explicit matrix calculation reveals that this is precisely the same set as in (iii).

Now we can prove (5.2.1). Let M and N be as in (ii). Choose a basis $n_1, \dots, n_b$ for N and, for each $\varepsilon \in X_m$, a basis $m_{\varepsilon,1}, \dots, m_{\varepsilon,a_\varepsilon}$ for M_ε. Also let $\{z_\zeta \mid \zeta \in X_n\}$ be a basis of the semisimple group algebra FW_n such that $wz_\zeta = \zeta(w)z_\zeta$ for each $w \in W_n$. Using (5.2c)(i) and the definition of induced module, we can write down a basis of $M \diamond_r N = FH_{m+n} \otimes_{FR_{m,n}} (M \otimes N)$ as follows:

$$\{uw_{\underline{i}}z_\zeta \otimes m_{\varepsilon,j} \otimes n_k \mid \underline{i} \in \Omega_{m,n}, u \in U_{\underline{i}}, \varepsilon \in X_m, \zeta \in X_n, 1 \le j \le a_\varepsilon, 1 \le k \le b\}.$$

The basis vector $uw_{\underline{i}}z_\zeta \otimes m_{\varepsilon,j} \otimes n_k$ has weight $uw_{\underline{i}}(\zeta + \varepsilon)$. Now, for $\underline{i} \in \Omega_{m,n}, u \in U_{\underline{i}}$, we have that $w_{\underline{i}}^{-1}u^{-1}\varepsilon_{m+n} = w_{\underline{i}}^{-1}\varepsilon_{m+n}$ which either equals ε_{m+n} if $i_{m+n} = m+n$, or ε_m if $i_m = m+n$. Hence:

(5.2d) *For* $\underline{i} \in \Omega_{m,n}, u \in U_{\underline{i}}, \zeta \in X_n, \varepsilon \in X_m$, *we have that* $uw_{\underline{i}}(\zeta + \varepsilon) = \varepsilon_{m+n}$ *if and only if* either $\varepsilon = 0, \zeta = \varepsilon_{m+n}, i_{m+n} = m+n$ or $\varepsilon = \varepsilon_m, \zeta = 0, i_m = m+n$.

We deduce that the ε_{m+n}-weight space of $M \diamond_r N$ splits as a direct sum of H_{m+n-1}-modules, where the first has basis

$$\left\{(uw_{\underline{i}})z_{\varepsilon_{m+n}} \otimes m_{0,j} \otimes n_k \;\middle|\; \begin{array}{l} \underline{i} \in \Omega_{m,n}, i_{m+n} = m+n, u \in U_{\underline{i}}, \\ 1 \le j \le a_0, 1 \le k \le b \end{array}\right\} \tag{5.2.2}$$

and the second has basis

$$\left\{(uw_{\underline{i}}\pi^{-1})\pi z_0 \otimes m_{\varepsilon_m,j} \otimes n_k \;\middle|\; \begin{array}{l} \underline{i} \in \Omega_{m,n}, i_m = m+n, u \in U_{\underline{i}}, \\ 1 \le j \le a_{\varepsilon_m}, 1 \le k \le b \end{array}\right\}. \tag{5.2.3}$$

Now, the $z_{\varepsilon_{m+n}} \otimes m_{0,j} \otimes n_k$ span an $FQ_{m,n-1}$-submodule of $M \diamond_r N$ isomorphic to the module obtained by inflating $(f_0 M) \boxtimes (N \downarrow H_{n-1})$ along the natural quotient map $Q_{m,n-1} \to G_m \times H_{n-1}$. Combining this with (5.2c)(ii), the definition of the operator $\diamond_\ell$ and the characterization of induced modules, we see that the vectors (5.2.2) span a module isomorphic to $(f_0 M) \diamond_\ell (N \downarrow_{H_{n-1}})$. Similarly, the $\pi z_0 \otimes m_{\varepsilon_m,j} \otimes n_k$ span an $FR_{m-1,n}$-submodule of $M \diamond_r N$ isomorphic to the module obtained by inflating $(f_+ M) \boxtimes N$ along the natural quotient map $R_{m-1,n} \to H_{m-1} \times G_n$. So using (5.2c)(iii) and the characterization of induced modules, we see that the vectors in (5.2.3) span a module isomorphic to $(f_+ M) \diamond_r N$. □

5.2e. **Corollary.** (i) *Given an* FG_m*-module* M *and an* FH_n*-module* N,

$$f_i(M \diamond_\ell N) \cong M \diamond (f_i N).$$

(ii) *Given an* FH_m*-module* M *and an* FG_n*-module* N,

$$f_i(M \diamond_r N) \cong (f_i M) \diamond N \oplus \bigoplus_{j=0}^{i-1} (f_j M) \diamond (f_{i-j-1}(N \downarrow_{H_{n-1}})).$$

Proof. Use the corresponding parts of Theorem 5.2b and induction on i. □

5.2f. **Corollary.** *Suppose that for $i = 1, 2, 3$, we are given an FG_{n_i}-module M_i and an FH_{n_i}-module N_i. Then, there are isomorphisms:*

$$M_1 \diamond (M_2 \diamond M_3) \cong (M_1 \diamond M_2) \diamond M_3, \tag{5.2.4}$$

$$N_1 \diamond_r (M_2 \diamond M_3) \cong (N_1 \diamond_r M_2) \diamond_r M_3, \tag{5.2.5}$$

$$M_1 \diamond_\ell (N_2 \diamond_r M_3) \cong (M_1 \diamond_\ell N_2) \diamond_r M_3, \tag{5.2.6}$$

$$(M_1 \diamond M_2) \diamond_\ell N_3 \cong M_1 \diamond_\ell (M_2 \diamond_\ell N_3). \tag{5.2.7}$$

Proof. Of these, (5.2.4) is a consequence of associativity of Harish-Chandra induction (2.2b). The remaining formulae follow in very similar ways from (5.2.4) and Corollary 5.2e, so we just prove (5.2.5) (the hardest) to illustrate the argument. In view of Lemma 5.1a, it suffices to check that for each $i \geq 0$,

$$f_i(N_1 \diamond_r (M_2 \diamond M_3)) \cong f_i((N_1 \diamond_r M_2) \diamond_r M_3).$$

One then expands both sides using (5.2.4), Corollary 5.2e and Lemma 5.2a; in both cases, one obtains the formula

$$(f_i N_1) \diamond M_2 \diamond M_3 \oplus \bigoplus_{j=0}^{i-1} \bigoplus_{k=0}^{i-j-2} (f_j N_1) \diamond f_k(M_2 \downarrow_{H_{n_2-1}}) \diamond f_{i-j-k-2}(M_3 \downarrow_{H_{n_3-1}})$$
$$\oplus \bigoplus_{j=0}^{i-1} (f_j N_1) \diamond M_2 \diamond f_{i-j-1}(M_3 \downarrow_{H_{n_3-1}})$$
$$\oplus \bigoplus_{j=0}^{i-1} (f_j N_1) \diamond f_{i-j-1}(M_2 \downarrow_{H_{n_2}-1}) \diamond M_3.$$

Hence, the two are isomorphic. □

In view of Corollary 5.2f, any multiple product involving the operators $\diamond, \diamond_r$ and $\diamond_\ell$ can always be rewritten as an expression containing the operators $\diamond_r$ and $\diamond_\ell$ *at most once*, and there is never any ambiguity over brackets.

5.2g. **Corollary.** *Let M be an FG_m-module and N be an FG_n-module. Write $M \uparrow^{H_m} = \bigoplus_{i \geq 0} e_i M_i$ and $N \uparrow^{H_n} = \bigoplus_{j \geq 0} e_j N_j$. Then:*

$$(M \diamond N) \uparrow^{H_{m+n}} \cong \bigoplus_{i,j \geq 0} e_{i+j}(M_i \diamond N_j), \tag{5.2.8}$$

$$(M \diamond N) \downarrow_{H_{m+n-1}} \cong \bigoplus_{\substack{i,j \geq 0 \\ (i,j) \neq (0,0)}} e_{i+j-1}(M_i \diamond N_j). \tag{5.2.9}$$

Proof. We prove (5.2.9); (5.2.8) then follows immediately by Lemma 5.1c(ii). It suffices by Lemma 5.1a(ii) and Lemma 5.2a to show that for all $k \geq 0$,

$$f_k((M \downarrow_{H_{m-1}}) \diamond_r N) \oplus M \diamond_\ell (N \downarrow_{H_{n-1}})) \cong \bigoplus_{i,j \geq 0, i+j-1=k} M_i \diamond N_j.$$

Note that $M = M_0$ and $M \downarrow_{H_{m-1}} \cong \bigoplus_{i \geq 0} e_i M_{i+1}$, and similarly for N, thanks to Lemma 5.1c. So, using in addition Corollary 5.2e, we see

$$f_k((M \downarrow_{H_{m-1}}) \diamond_r N) \oplus M \diamond_\ell (N \downarrow_{H_{n-1}})) \cong M_0 \diamond N_{k+1} \oplus M_{k+1} \diamond N_0 \oplus \bigoplus_{j=0}^{k-1} M_{j+1} \diamond N_{k-j}.$$

The proof follows. □

5.3. The affine cuspidal algebra

Now we introduce an affine analogue of the cuspidal algebra. Fix $\sigma \in \bar{\mathbb{F}}_q^\times$ of degree d over $\mathbb{F}_q$ and $k \geq 0$. Set $n = kd$ throughout the section. Recall the definition of the FG_n-module $M^k(\sigma)$ from (2.4.2); in the case $k = 0$, this is the trivial module over the trivial group G_0. Note that for each $0 \leq j \leq k$, the module $e_{dj} M^{k-j}(\sigma)$ is an FH_n-module. Define the *affine cuspidal algebra*

$$D_{\sigma,k} = D_{F,(\sigma)^k}(GL_n(\mathbb{F}_q)) = FH_n \Big/ \bigcap_{j=0}^{k} \mathrm{ann}_{FH_n}(e_{dj} M^{k-j}(\sigma)).$$

So $D_{\sigma,k}$ can be thought of as the image of FH_n under the representation afforded by the module $\bigoplus_{j=0}^k e_{dj} M^{k-j}(\sigma)$. The following lemma should serve as motivation for this definition:

5.3a. **Lemma.** $M^k(\sigma) \uparrow^{H_n} \cong \bigoplus_{j=0}^{k} \binom{k}{j} e_{dj} M^{k-j}(\sigma).$

Proof. Use induction on k. The case $k = 1$ follows from Corollary 5.1f(i). For the induction step, write $M^k(\sigma) = M^{k-1}(\sigma) \diamond M(\sigma)$ and apply Corollary 5.2g. □

5.3b. **Corollary.** *If M is a $C_{\sigma,k}$-module, then $M \uparrow^{H_n}$ is a $D_{\sigma,k}$-module.*

Proof. It suffices to check this on projective indecomposables. In turn, since every projective indecomposable $C_{\sigma,k}$-module is a submodule of $M^k(\sigma)$ according to Theorem 3.4g, we just need to check this in the special case $M = M^k(\sigma)$. But in that case, the corollary follows from Lemma 5.3a and the definition of $D_{\sigma,k}$. □

Now fix $h \geq k$ for the remainder of the chapter. Write $S_{h,k}$ for the q^d-Schur algebra $S_{F,q^d}(h,k)$ as in §1.2. We need to work now with the algebra

$$S_{h,\leq k} = \bigoplus_{i=0}^{k} S_{h,i}. \qquad (5.3.1)$$

This is a quasi-hereditary algebra with weight poset $\Lambda^+(h, \leq k) = \bigcup_{i=0}^k \Lambda^+(h,i)$, partially ordered by the union of the dominance orders on each $\Lambda^+(h,i)$. For

$\lambda \in \Lambda^+(h, \leq k)$, we write $L_h(\lambda)$, $\Delta_h(\lambda)$, etc... for the usual modules over $S_{h,\leq k}$ corresponding to the partition λ. So if $\lambda \in \Lambda^+(h, i)$ then these are the usual modules over the summand $S_{h,i}$ of $S_{h,\leq k}$, with all other summands acting as zero.

Also, for any $0 \leq j \leq k$, define

$$Z_j = \bigoplus_{\nu \in \Lambda(h,k-j)} \dot{Z}^\nu(\sigma), \tag{5.3.2}$$

which is an $(FG_{n-dj}, S_{h,k-j})$-bimodule. Applying the functor e_{dj}, we obtain an $(FH_n, S_{h,k-j})$-bimodule $e_{dj}Z_j$. Hence, the module

$$\widehat{Z} = \bigoplus_{j=0}^{k} e_{dj} Z_j \tag{5.3.3}$$

is naturally an $(FH_n, S_{h,\leq k})$-bimodule.

5.3c. **Theorem.** *With notation as above, $\widehat{Z}$ is a $D_{\sigma,k}$-module which is a projective generator for* $\operatorname{mod}(D_{\sigma,k})$, *and the endomorphism algebra* $\operatorname{End}_{D_{\sigma,k}}(\widehat{Z})$ *is precisely the algebra $S_{h,\leq k}$.*

Proof. Let us first recall some known facts from Theorem 3.4g. First, for each $0 \leq j \leq k$, $S_{h,k-j}$ is the endomorphism algebra $\operatorname{End}_{FG_{n-dj}}(Z_j)$. Second, Z_j contains a copy of each of its composition factors in its head. Finally, as explained in the proof of Theorem 3.4g, Lemma 3.4f gives that

$$\dim \operatorname{Hom}_{FG_{n-dj}}(P_j, Z_j) = \dim \operatorname{Hom}_{FG_{n-dj}}(Z_j, Z_j)$$

where P_j is the projective cover of Z_j in the category $\operatorname{mod}(FG_{n-dj})$.

Now, for $0 \leq j \leq k$, we regard Z_j (resp. P_j) as an $FG_n \oplus FG_{n-1} \oplus \cdots \oplus FG_0$-module so that the summand FG_{n-dj} acts on Z_j (resp. P_j) as given and all other summands act as zero. Then, applying the Morita equivalence $e_0 \oplus \cdots \oplus e_n$ of Lemma 5.1a(ii) to the statements in the previous paragraph, we deduce immediately that:

(i) $\operatorname{End}_{FH_n}(\widehat{Z}) = S_{h,\leq k}$;

(ii) $\widehat{Z}$ contains a copy of every composition factor in its head;

(iii) $\dim \operatorname{Hom}_{FH_n}(\widehat{P}, \widehat{Z}) = \dim \operatorname{Hom}_{FH_n}(\widehat{Z}, \widehat{Z})$, where $\widehat{P}$ is the projective cover of $\widehat{Z}$ in the category $\operatorname{mod}(FH_n)$.

Now we prove the theorem. Since each Z_j is a direct sum of submodules of $M^{k-j}(\sigma)$ by definition, $e_{dj}Z_j$ is a direct sum of submodules of $e_{dj}M^{k-j}(\sigma)$. Hence, $\operatorname{ann}_{FH_n}(e_{dj}M^{k-j}(\sigma))$ annihilates $e_{dj}Z_j$. This shows that the action of FH_n on $\widehat{Z}$ factors through the quotient $D_{\sigma,k}$ to induce a well-defined $D_{\sigma,k}$-module.

Moreover, $\widehat{Z}$ contains a summand isomorphic to $\bigoplus_{j=0}^{k} e_{dj}M^{k-j}(\sigma)$, hence $\widehat{Z}$ is a faithful $D_{\sigma,k}$-module. The endomorphism algebra $\operatorname{End}_{D_{\sigma,k}}(\widehat{Z})$ is just $\operatorname{End}_{FH_n}(\widehat{Z})$, hence isomorphic to $S_{h,\leq k}$ by (i). Using (iii), Lemma 3.2a and faithfulness, $\widehat{Z}$ is a projective $D_{\sigma,k}$-module. Finally, it is a projective generator for $\text{mod}(D_{\sigma,k})$ by (ii) and faithfulness, using the same argument as in the last paragraph of the proof of Theorem 3.4g. $\square$

Now we proceed as we did in §3.5. Introduce functors

$$\alpha_{\sigma,h,\leq k} : \text{mod}(D_{\sigma,k}) \to \text{mod}(S_{h,\leq k}), \qquad \alpha_{\sigma,h,\leq k} = \text{Hom}_{D_{\sigma,k}}(\widehat{Z}, ?), \tag{5.3.4}$$

$$\beta_{\sigma,h,\leq k} : \text{mod}(S_{h,\leq k}) \to \text{mod}(D_{\sigma,k}), \qquad \beta_{\sigma,h,\leq k} = \widehat{Z} \otimes_{S_{h,\leq k}} ?. \tag{5.3.5}$$

Theorem 5.3c immediately implies that:

(5.3d) *The functors $\alpha_{\sigma,h,\leq k}$ and $\beta_{\sigma,h,\leq k}$ are mutually inverse equivalences of categories between $\text{mod}(D_{\sigma,k})$ and $\text{mod}(S_{h,\leq k})$.*

The next lemma identifies the various standard modules for $D_{\sigma,k}$ coming from the above Morita equivalence:

5.3e. **Lemma.** *For any $0 \leq j \leq k$ and $\lambda \vdash (k-j)$,*

$$\begin{aligned} e_{dj} L(\sigma, \lambda) &\cong \beta_{\sigma,h,\leq k}(L_h(\lambda')), \\ e_{dj} \Delta(\sigma, \lambda) &\cong \beta_{\sigma,h,\leq k}(\Delta_h(\lambda')), \\ e_{dj} \nabla(\sigma, \lambda) &\cong \beta_{\sigma,h,\leq k}(\nabla_h(\lambda')), \\ e_{dj} T(\sigma, \lambda) &\cong \beta_{\sigma,h,\leq k}(T_h(\lambda')). \end{aligned}$$

Proof. We prove something more general. For each $0 \leq j \leq k$, the algebra $S_{h,k-j}$ is a summand of $S_{h,\leq k}$. So there is a natural inflation functor

$$\text{infl}_j : \text{mod}(S_{h,k-j}) \to \text{mod}(S_{h,\leq k}).$$

The lemma follows immediately from the claim:

(5.3f) *The following functors are isomorphic:*

$$\begin{aligned} e_{dj} \circ \beta_{\sigma,h,k-j} &: \text{mod}(S_{h,k-j}) \to \text{mod}(FH_n), \\ \beta_{\sigma,h,\leq k} \circ \text{infl}_j &: \text{mod}(S_{h,k-j}) \to \text{mod}(FH_n). \end{aligned}$$

To prove (5.3f), let $i_j \in S_{h,\leq k}$ be the central idempotent corresponding to the identity of the summand $S_{h,k-j}$ of $S_{h,\leq k}$. So, $S_{h,k-j} = i_j S_{h,\leq k} i_j$, and the functor infl_j is the functor $S_{h,\leq k} i_j \otimes_{S_{h,k-j}} ?$. Also observe that $\widehat{Z} i_j = e_{dj} Z_j$, so there is an isomorphism of functors

$$\beta_{\sigma,h,\leq k} \circ \text{infl}_j = \widehat{Z} \otimes_{S_{h,\leq k}} (S_{h,\leq k} i_j \otimes_{S_{h,k-j}} ?) \cong \widehat{Z} i_j \otimes_{S_{h,k-j}} ? = (e_{dj} Z_j) \otimes_{S_{h,k-j}} ?.$$

Now use Lemma 5.1d and associativity of tensor product to deduce that

$$(e_{dj} Z_j) \otimes_{S_{h,k-j}} ? \cong e_{dj} \circ (Z_j \otimes_{S_{h,k-j}} ?) = e_{dj} \circ \beta_{\sigma,h,k-j},$$

completing the proof of (5.3f). □

To summarize: the algebra $D_{\sigma,k}$ is a quasi-hereditary algebra with weight poset $\{\lambda \vdash (k-j) \,|\, 0 \leq j \leq k\}$ partially ordered by $\geq$ (the opposite order to $S_{h,\leq k}$ since we have transposed partitions). Moreover, the $\{e_{dj}L(\sigma,\lambda)\}$, $\{e_{dj}\Delta(\sigma,\lambda)\}$, $\{e_{dj}\nabla(\sigma,\lambda)\}$ and $\{e_{dj}T(\sigma,\lambda)\}$ for all $0 \leq j \leq k, \lambda \vdash (k-j)$ give the irreducible, standard, costandard and indecomposable tilting $D_{\sigma,k}$-modules.

5.4. The branching rule from GL_n to AGL_{n-1}

According to Lemma 5.1c(ii), understanding induction from G_n to H_n or restriction from G_n to H_{n-1} are essentially equivalent problems. It turns out that it is more convenient to study induction first and then deduce the consequences for restriction. Continue with the notation set up in the previous section. Also, let $S_{h+1,k}$ denote the q^d-Schur algebra $S_{F,q^d}(h+1,k)$ and Z denote the $(FG_n, S_{h+1,k})$-bimodule

$$Z = \bigoplus_{\nu \in \Lambda(h+1,k)} \dot{Z}^{\nu}(\sigma).$$

So Z is a projective generator for $\mathrm{mod}(C_{\sigma,k})$. Now, the Levi subalgebra $S_{\nu,k}$ of (1.3g), in the special case $\nu = (h,1)$, is a subalgebra of $S_{h+1,k}$, isomorphic by (1.3.8) to

$$\bigoplus_{j=0}^{k} S_{h,k-j} \otimes S_{1,j}.$$

For any j, the algebra $S_{1,j}$ is one dimensional, so we can identify this subalgebra with the algebra $S_{h,\leq k}$ of (5.3.1). To be explicit, for $\mu \in \Lambda(h,k-j)$, write $\mu[j]$ for the $(h+1)$-tuple in $\Lambda(h+1,k)$ with $(h+1)$-entry equal to j, and all other entries being the same as in the h-tuple μ. Also view Σ_{k-j} as the naturally embedded subgroup of Σ_k. Then:

(5.4a) *The embedding $S_{h,\leq k} \hookrightarrow S_{h+1,k}$ maps the standard basis element $\phi^u_{\mu,\lambda}$ of $S_{h,k-j} \subset S_{h,\leq k}$, for $\mu,\lambda \in \Lambda(h,k-j), u \in D_{\mu,\lambda}$ to the standard basis element $\phi^{u'}_{\mu[j],\lambda[j]}$ of $S_{h+1,k}$, where u' is the image of u under the natural embedding $\Sigma_{k-j} \hookrightarrow \Sigma_k$.*

For the next lemma, we regard the restriction $\widehat{Z}\downarrow_{G_n}$ of the $(FH_n, S_{h,\leq k})$-bimodule $\widehat{Z}$ of (5.3.3) as an $(FG_n, S_{h,\leq k})$-bimodule in the natural way.

5.4b. **Lemma.** *There is a surjective $(FG_n, S_{h,\leq k})$-bimodule homomorphism*

$$\theta : \widehat{Z}\downarrow_{G_n} \to Z,$$

such that all other FG_n-homomorphisms $\widehat{Z}\downarrow_{G_n} \to Z$ factor through θ.

Proof. Consider

$$Z' = \bigoplus_{j=0}^{k} Z_j \diamond Z^j(\sigma), \tag{5.4.1}$$

where Z_j is as in (5.3.2). Each Z_j is an $(FG_{n-dj}, S_{h,k-j})$-bimodule, so the summand $Z_j \diamond Z^j(\sigma)$ of Z' is an $(FG_n, S_{h,k-j})$-bimodule (we have applied the functor $? \diamond Z^j(\sigma)$). This makes Z' into an $(FG_n, S_{h,\leq k})$-bimodule. Now let $\omega : Z' \to Z$ be the evident FG_n-module isomorphism that identifies the summand $\dot{Z}^\nu(\sigma) \diamond Z^j(\sigma)$ of Z', for any $0 \leq j \leq k$ and $\nu \in \Lambda(h, k-j)$, with the summand $\dot{Z}^{\nu[j]}(\sigma)$ of Z. We claim that ω is a right $S_{h,\leq k}$-module homomorphism, hence an isomorphism of $(FG_n, S_{h,\leq k})$-bimodules.

For the claim, we consider a standard basis element $\phi^u_{\mu,\lambda}$ of the summand $S_{h,k-j}$ of $S_{h,\leq k}$, for $\mu, \lambda \in \Lambda(h, k-j)$ and $u \in D_{\mu,\lambda}$. By (5.4a), this coincides with the element $\phi^{u'}_{\mu[j],\lambda[j]}$ of $S_{h+1,k}$, so by Theorem 3.4c it acts as zero on all summands of Z except for the ones of the form $\dot{Z}^{\nu[j]}(\sigma)$ for $\nu \in \Lambda(h, k-j)$. Moreover, the action of $\phi^u_{\mu[j],\lambda[j]}$ on such a summand $\dot{Z}^{\nu[j]}(\sigma)$ is zero unless $\nu = \mu$, in which case the action is induced by right multiplication in $M^k(\sigma)$ by the element $\sum_{w \in \Sigma_{\mu[j]} u' \Sigma_{\lambda[j]} \cap D^{-1}_{\mu[j]}} T^\#_w \in H_k$. On the other hand, the action of the element $\phi^u_{\mu,\lambda}$ on Z' is zero on all summands of Z' except $Z_j \diamond Z^j(\sigma)$, while on any summand $\dot{Z}^\nu(\sigma) \diamond Z^j(\sigma)$ of $Z_j \diamond Z^j(\sigma)$ the action of $\phi^u_{\mu,\lambda}$ is induced by applying the functor $? \diamond Z^j(\sigma)$ to the action coming from right multiplication in $M^{k-j}(\sigma)$ by the element $\sum_{w \in \Sigma_\mu u \Sigma_\lambda \cap D^{-1}_\mu} T^\#_w \in H_{k-j}$. To prove the claim, it just remains to observe that the element $\sum_{w \in \Sigma_{\mu[j]} u' \Sigma_{\lambda[j]} \cap D^{-1}_{\mu[j]}} T^\#_w$ of H_k is the same as the element $\sum_{w \in \Sigma_\mu u \Sigma_\lambda \cap D^{-1}_\mu} T^\#_w$ of the naturally embedded subalgebra $H_{k-j} \subset H_{k-j} \otimes H_j \subset H_k$. So the two actions do indeed agree under the bijection ω because of Lemma 3.2f(i).

Now we can prove the lemma. For each $0 \leq j \leq k$, fix a non-zero surjection $\theta_j : \Gamma_{dj} \to Z^j(\sigma)$, which exists and is unique up to scalars by Lemma 3.4d(i) and (3.4h). According to Theorem 5.1e, the $(FG_n, S_{h,\leq k})$-bimodule $\widehat{Z} \downarrow_{G_n}$ can be identified with the bimodule $\bigoplus_{j=0}^k Z_j \diamond \Gamma_{dj}$. Also, by the claim, we can use the map ω to identify the $(FG_n, S_{h,\leq k})$-bimodule Z with $\bigoplus_{j=0}^k Z_j \diamond Z^j(\sigma)$. Now define $\theta = \bigoplus_{j=0}^k \mathrm{id}_{Z_j} \otimes \phi_j$, to give the desired a surjective $(FG_n, S_{h,\leq k})$-bimodule homomorphism

$$\theta : \widehat{Z} \downarrow_{G_n} = \bigoplus_{j=0}^{k} Z_j \diamond \Gamma_{dj} \to \bigoplus_{j=0}^{k} Z_j \diamond Z^j(\sigma) = Z.$$

It just remains to check that any other FG_n-homomorphism $\bigoplus_{j=0}^k Z_j \diamond \Gamma_{dj} \to \bigoplus_{j=0}^k Z_j \diamond Z^j(\sigma)$ factors through θ. Expanding the direct sums in the definition of Z_j, this follows if we can show that

$$\dim \mathrm{Hom}_{FG_n}(\dot{Z}^\nu(\sigma) \diamond \Gamma_{dj}, \dot{Z}^\mu(\sigma)) = \dim \mathrm{Hom}_{FG_n}(\dot{Z}^\nu(\sigma) \diamond Z^j(\sigma), \dot{Z}^\mu(\sigma))$$

for every $\nu \in \Lambda(h, k-j), \mu \in \Lambda(h+1, k)$. But given (3.4h), this follows at once from the stronger statement of Lemma 3.4f. $\square$

We remark that $C_{\sigma,k}$ is not a subalgebra of $D_{\sigma,k}$ in any natural way. However, by Corollary 5.3b, we can still define a functor

$$\mathrm{ind}_{C_{\sigma,k}}^{D_{\sigma,k}} : \mathrm{mod}(C_{\sigma,k}) \to \mathrm{mod}(D_{\sigma,k}),$$

namely, the restriction of the functor $\mathrm{ind}_{G_n}^{H_n} : \mathrm{mod}(FG_n) \to \mathrm{mod}(FH_n)$ to the full subcategories $\mathrm{mod}(C_{\sigma,k}) \subset \mathrm{mod}(FG_n)$ and $\mathrm{mod}(D_{\sigma,k}) \subset \mathrm{mod}(FH_n)$. The following theorem relates this induction functor to the restriction functor from $S_{h+1,k}$ to the Levi subalgebra $S_{h,\leq k}$:

5.4c. **Theorem.** *The following functors are isomorphic:*

$$\mathrm{ind}_{C_{\sigma,k}}^{D_{\sigma,k}} \circ \beta_{\sigma,h+1,k} : \mathrm{mod}(S_{h+1,k}) \to \mathrm{mod}(D_{\sigma,k}),$$
$$\beta_{\sigma,h,\leq k} \circ \mathrm{res}_{S_{h,\leq k}}^{S_{h+1,k}} : \mathrm{mod}(S_{h+1,k}) \to \mathrm{mod}(D_{\sigma,k}).$$

Proof. Since Z is an $(FG_n, S_{h+1,k})$-bimodule, we can view $\mathrm{Hom}_{FG_n}(Z, Z)$ as an $(S_{h+1,k}, S_{h+1,k})$-bimodule, where the actions are defined by $(s\psi)(z) = \psi(zs)$ and $(\psi s)(z) = \psi(z)s$ for $s \in S_{h+1,k}, z \in Z$ and $\psi : Z \to Z$. Then we can restrict the left action to the subalgebra $S_{h,\leq k} \subset S_{h+1,k}$ to view $\mathrm{Hom}_{FG_n}(Z, Z)$ an $(S_{h,\leq k}, S_{h+1,k})$-bimodule. We first show that

$$\phi : \mathrm{Hom}_{FG_n}(Z, Z) \xrightarrow{\sim} \mathrm{Hom}_{FG_n}(\widehat{Z} \downarrow_{G_n}, Z), \qquad f \mapsto f \circ \theta,$$

is an $(S_{h,\leq k}, S_{h+1,k})$-bimodule isomorphism, where θ is the surjection of Lemma 5.4b. Well, ϕ is bijective since θ is a surjection and every FG_n-homomorphism $\widehat{Z} \downarrow_{G_n} \to Z$ factors through θ. It is obviously a right $S_{h+1,k}$-module map, and a routine check using the fact that θ is $S_{h,\leq k}$-equivariant shows that it is a left $S_{h,\leq k}$-module map.

Now note that the $(S_{h+1,k}, S_{h+1,k})$-bimodule $\mathrm{Hom}_{FG_n}(Z, Z)$ is just the regular bimodule $S_{h+1,k}$ itself, in view of Theorem 3.4c. So, viewing $S_{h+1,k}$ as an $(S_{h,\leq k}, S_{h+1,k})$-bimodule by restricting the left action, we have shown that the bimodules $\mathrm{Hom}_{FG_n}(\widehat{Z} \downarrow_{G_n}, Z)$ and $S_{h+1,k}$ are isomorphic. Hence, by Frobenius reciprocity, there is an isomorphism $S_{h+1,k} \cong \mathrm{Hom}_{FG_n}(\widehat{Z}, Z \uparrow^{H_n})$ as $(S_{h,\leq k}, S_{h+1,k})$-bimodules. Since Z is a $C_{\sigma,k}$-module, we even have by Corollary 5.3b that

$$S_{h+1,k} \cong \mathrm{Hom}_{D_{\sigma,k}}(\widehat{Z}, \mathrm{ind}_{C_{\sigma,k}}^{D_{\sigma,k}} Z) \tag{5.4.2}$$

as $(S_{h,\leq k}, S_{h+1,k})$-bimodules.

Now regard the functors $\alpha_{\sigma,h,\leq k}$ and $\beta_{\sigma,j,\leq k}$ of (5.3.4)–(5.3.5) as bimodule functors in the obvious way to obtain functors:

$$\alpha_{\sigma,h,\leq k} : \mathrm{bimod}(D_{\sigma,k}, S_{h+1,k}) \to \mathrm{bimod}(S_{h,\leq k}, S_{h+1,k}),$$
$$\beta_{\sigma,h,\leq k} : \mathrm{bimod}(S_{h,\leq k}, S_{h+1,k}) \to \mathrm{bimod}(D_{\sigma,k}, S_{h+1,k}).$$

By (5.3d) and naturality, $\beta_{\sigma,h,\leq k} \circ \alpha_{\sigma,h,\leq k}$ is isomorphic to the identity functor on $\mathrm{bimod}(D_{\sigma,k}, S_{h+1,k})$. So if we apply $\beta_{\sigma,h,\leq k}$ to (5.4.2) we deduce at once that there is a $(D_{\sigma,k}, S_{h+1,k})$-bimodule isomorphism

$$\widehat{Z} \otimes_{S_{h,\leq k}} S_{h+1,k} \cong \mathrm{ind}_{C_{\sigma,k}}^{D_{\sigma,k}} Z. \tag{5.4.3}$$

Now we can prove the theorem. For any $S_{h+1,k}$-module M, (5.4.3) and associativity of tensor products gives us natural isomorphisms:

$$\begin{aligned}
\beta_{\sigma,h,\leq k} \circ \mathrm{res}_{S_{h,\leq k}}^{S_{h+1,k}}(M) &= \widehat{Z} \otimes_{S_{h,\leq k}} (\mathrm{res}_{S_{h,\leq k}}^{S_{h+1,k}} M) \\
&\cong \widehat{Z} \otimes_{S_{h,\leq k}} (S_{h+1,k} \otimes_{S_{h+1,k}} M) \\
&\cong (\widehat{Z} \otimes_{S_{h,\leq k}} S_{h+1,k}) \otimes_{S_{h+1,k}} M \\
&\cong (\mathrm{ind}_{C_{\sigma,k}}^{D_{\sigma,k}} Z) \otimes_{S_{h+1,k}} M \\
&\cong \mathrm{ind}_{C_{\sigma,k}}^{D_{\sigma,k}} (Z \otimes_{S_{h+1,k}} M) = \mathrm{ind}_{C_{\sigma,k}}^{D_{\sigma,k}} \circ \beta_{\sigma,h,k}(M).
\end{aligned}$$

Hence, the functors $\beta_{\sigma,h,\leq k} \circ \mathrm{res}_{S_{h,\leq k}}^{S_{h+1,k}}$ and $\mathrm{ind}_{C_{\sigma,k}}^{D_{\sigma,k}} \circ \beta_{\sigma,h,k}$ are isomorphic. □

Using Theorem 5.4c, we obtain the following description of composition multiplicities:

5.4d. **Corollary.** *Suppose that* $\lambda \vdash k$.

(i) *All composition factors of* $L(\sigma,\lambda)\uparrow^{H_n}$ *are of the form* $e_{dj}L(\sigma,\mu)$ *for* $0 \leq j \leq k$ *and* $\mu \vdash (k-j)$. *Moreover, for such* μ, j,

$$[L(\sigma,\lambda)\uparrow^{H_n} : e_{dj}L(\sigma,\mu)] = [\mathrm{res}_{S_{h,\leq k}}^{S_{h+1,k}} L_{h+1}(\lambda') : L_h(\mu')].$$

(ii) *All composition factors of* $L(\sigma,\lambda)\downarrow_{H_{n-1}}$ *are of the form* $e_{dj-1}L(\sigma,\mu)$ *for* $1 \leq j \leq k$ *and* $\mu \vdash (k-j)$. *Moreover, for such* μ, j,

$$[L(\sigma,\lambda)\downarrow_{H_{n-1}} : e_{dj-1}L(\sigma,\mu)] = [\mathrm{res}_{S_{h,\leq k}}^{S_{h+1,k}} L_{h+1}(\lambda') : L_h(\mu')].$$

Proof. (i) By the theorem, (3.5.3) and the definition of $\mathrm{ind}_{C_{\sigma,k}}^{D_{\sigma,k}}$,

$$L(\sigma,\lambda)\uparrow^{H_n} \cong \mathrm{ind}_{C_{\sigma,k}}^{D_{\sigma,k}} \circ \beta_{\sigma,h+1,k}(L_{h+1}(\lambda')) \cong \beta_{\sigma,h,\leq k} \circ \mathrm{res}_{S_{h,\leq k}}^{S_{h+1,k}}(L_{h+1}(\lambda')).$$

Now the result follows immediately from Lemma 5.3e.

(ii) Apply Lemma 5.1c(i) to (i). □

5.4e. **Remarks.** (i) Corollary 5.4d can be generalized easily to give a formula for the composition multiplicities in $L\uparrow^{H_n}$ (resp. $L\downarrow_{H_{n-1}}$) for an arbitrary irreducible FG_n-module of the form (4.4.1), using Corollary 5.2g and induction.

(ii) Theorem 5.4c can also be used to reduce other natural questions about the structure of $L(\sigma,\lambda)\downarrow_{H_{n-1}}$ to analogous problems about $L_{h+1}(\lambda')\downarrow_{S_{h,\leq k}}$. The latter can in turn be answered by studying the branching rule from quantum GL_{h+1} to quantum GL_h. For instance, the problem of determining the socle of $L(\sigma,\lambda)\downarrow_{H_{n-1}}$, or a criterion for complete reducibility of $L(\sigma,\lambda)\downarrow_{H_{n-1}}$, can be tackled in this way.

5.5. A dimension formula for irreducibles

In this section, we prove a dimension formula for irreducible FG_n-modules in terms of the characters of the irreducible $S_{h,k}$-modules (the latter being conjecturally determined by the Lusztig conjecture in many cases). Continue with the notation of the previous section. In addition, for $\lambda \in \Lambda^+(h+1,k), \mu \in \Lambda^+(h, \leq k)$ and $\nu \in \Lambda(h+1,k)$, we will write

$$b_{\lambda,\mu} = [\mathrm{res}^{S_{h+1,k}}_{S_{h,\leq k}} L_{h+1}(\lambda) : L_h(\mu)], \tag{5.5.1}$$

$$m_{\lambda,\nu} = \dim L_{h+1}(\lambda)_\nu, \tag{5.5.2}$$

i.e. $b_{\lambda,\mu}$ is the branching coefficient for restricting from the quantized enveloping algebra A_{h+1} to A_h, and $m_{\lambda,\nu}$ is the dimension of the ν-weight space of $L_{h+1}(\lambda)$.

We will need the following inductive description of the weight multiplicities:

5.5a. **Lemma.** *For $0 \leq j \leq k$, $\lambda \in \Lambda^+(h+1,k)$ and $\mu = (m_1, \dots, m_h) \in \Lambda(h, k-j)$,*

$$m_{\lambda,\mu[j]} = \sum_{\nu \in \Lambda^+(h,k-j)} b_{\lambda,\nu} m_{\nu,\mu}$$

where $\mu[j] = (m_1, \dots, m_h, j) \in \Lambda(h+1,k)$.

Proof. The restriction $\mathrm{res}^{S_{h+1,k}}_{S_{h,\leq k}} L_{h+1}(\lambda)$ splits as a direct sum of 'levels'

$$\mathrm{res}^{S_{h+1,k}}_{S_{h,\leq k}} L_{h+1}(\lambda) = \bigoplus_{j=0}^{k} L_{h+1}(\lambda)^{(j)},$$

where $L_{h+1}(\lambda)^{(j)}$ is a module for the summand $S_{h,k-j}$ of $S_{h,\leq k}$. Explicitly, $L_{h+1}(\lambda)^{(j)}$ is the sum of all weight spaces of $L_{h+1}(\lambda)$ corresponding to weights of the form $\nu = (n_1, \dots, n_{h+1}) \in \Lambda(h+1,k)$ with $n_{h+1} = j$. Equivalently, it is the largest submodule of $\mathrm{res}^{S_{h+1,k}}_{S_{h,\leq k}} L_{h+1}(\lambda)$ all of whose composition factors are of the form $L_h(\nu)$ for $\nu \in \Lambda^+(h, k-j)$. This means that $L_{h+1}(\lambda)^{(j)}$ has the same composition factors over $S_{h,k-j}$ as

$$\bigoplus_{\nu \in \Lambda^+(h,k-j)} b_{\lambda,\nu} L_h(\nu).$$

We see that the $\mu[j]$-weight space of $L_{h+1}(\lambda)$ has the same dimension as the μ-weight space of $\bigoplus_{\nu \in \Lambda^+(h,k-j)} b_{\lambda,\nu} L_h(\nu)$. The lemma follows. □

The next theorem describes of the restriction of the FG_n-module $L(\sigma, \lambda')$ to FG_{n-1}:

5.5b. **Theorem.** *For $\lambda \vdash k$,*

$$L(\sigma, \lambda') \downarrow_{G_{n-1}} \cong \bigoplus_{j=1}^{k} M_j \diamond \Gamma_{dj-1}$$

where M_j is an FG_{n-dj}-module having the same composition factors as the semisimple module $\bigoplus_{\mu \vdash (k-j)} b_{\lambda,\mu} L(\sigma, \mu')$.

Proof. By Corollary 5.4d(ii),

$$L(\sigma,\lambda')\downarrow_{H_{n-1}}\cong\bigoplus_{j=1}^{k} e_{dj-1}M_j.$$

Now apply Theorem 5.1e. □

5.5c. **Corollary.** *For* $\lambda \vdash k$,

$$\dim L(\sigma,\lambda')=\sum_{j=1}^{k}\left[\prod_{i=1}^{dj-1}(q^{n-i}-1)\right]\left[\sum_{\mu\vdash(k-j)} b_{\lambda,\mu}\dim L(\sigma,\mu')\right].$$

Proof. If M is an FG_m-module, then an elementary calculation shows that

$$\dim M\diamond\Gamma_n=\dim M\left[\prod_{i=1}^{n}(q^{m+i}-1)\right].$$

Now the corollary follows easily from Theorem 5.5b. □

Now we introduce some polynomials. For a composition $\mu=(m_1,m_2,\dots)\models n$, define the polynomial $R_\mu(t)\in\mathbb{Z}[t]$ by

$$R_\mu(t)=\prod_{i=1}^{n}(t^i-1)\Big/\prod_{i>0\text{ with }m_i>0}(t^{m_1+\dots+m_i}-1).$$

Let $\lambda=(l_1,\dots,l_a)$ be a partition of n of height a. We will write $\mu\sim\lambda$ if μ is any composition of height a obtained from λ by reordering the non-zero parts. For example, if $\lambda=(3,2,2)$, there are precisely three compositions μ with $\mu\sim\lambda$, namely, $\mu=(3,2,2),(2,3,2)$ and $(2,2,3)$. Define the polynomial $S_\lambda(t)\in\mathbb{Z}[t]$ by the formula

$$S_\lambda(t)=\sum_{\mu\sim\lambda}R_\mu(t). \tag{5.5.3}$$

For some very simple examples:

$$\begin{aligned}
S_{(1^n)}(t)&=1,\\
S_{(n)}(t)&=(t^{n-1}-1)(t^{n-2}-1)\dots(t-1),\\
S_{(d^k)}(t)&=\prod_{i=1}^{kd}(t^i-1)\Big/\prod_{i=1}^{k}(t^{di}-1),\\
S_{(2,1^{n-2})}(t)&=(t^n-1)/(t-1)-n.
\end{aligned}$$

We note in particular that $S_{(d^k)}(q)=|GL_{kd}(q):GL_k(q^d)|_{q'}$. Our main result is as follows:

5.5d. **Theorem.** *Suppose that $\sigma \in \bar{\mathbb{F}}_q^\times$ is of degree d over $\mathbb{F}_q$, and $n = kd$ for some $k \geq 1$. Then, for any $\lambda \vdash k$,*

$$\dim L(\sigma, \lambda') = S_{(d^k)}(q) \sum_{\mu \vdash k} m_{\lambda,\mu} S_\mu(q^d),$$

where $m_{\lambda,\mu}$ is as in (5.5.2).

Proof. For the proof, we will use the notation $\nu \Vdash k$ to mean that $\nu \sim \mu$ for some $\mu \vdash k$. So if $\nu \Vdash k$, ν is a composition of k with as many non-zero parts as its height. Now, if $\nu \sim \mu$, then $m_{\lambda,\mu} = m_{\lambda,\nu}$. So we can rewrite the statement we are trying to prove equivalently as

$$\dim L(\sigma, \lambda') = S_{(d^k)}(q) \sum_{\nu \Vdash k} m_{\lambda,\nu} R_\nu(q^d). \tag{5.5.4}$$

We will prove (5.5.4) by induction on k, the case $k = 1$ following from (2.4.1). For $k > 1$, we use Corollary 5.5c and the inductive hypothesis to obtain:

$$\begin{aligned} \dim L(\sigma, \lambda') &= \sum_{j=1}^{k} \left[\prod_{i=1}^{dj-1} (q^{n-i} - 1) \right] \sum_{\mu \vdash k-j} b_{\lambda,\mu} S_{(d^{k-j})}(q) \sum_{\nu \Vdash k-j} m_{\mu,\nu} R_\nu(q^d) \\ &= \sum_{j=1}^{k} \sum_{\nu \Vdash k-j} \left[S_{(d^{k-j})}(q) R_\nu(q^d) \prod_{i=1}^{dj-1} (q^{n-i} - 1) \right] \left[\sum_{\mu \vdash k-j} b_{\lambda,\mu} m_{\mu,\nu} \right]. \end{aligned}$$

Write $\nu[j]$ for the composition obtained from ν by replacing the first part equal to zero with j. Then, a calculation from the definitions shows that for $\nu \Vdash k - j$,

$$S_{(d^{k-j})}(q) R_\nu(q^d) \prod_{i=1}^{dj-1} (q^{n-i} - 1) = S_{(d^k)}(q) R_{\nu[j]}(q^d).$$

Also, Lemma 5.5a shows that

$$\sum_{\mu \vdash k-j} b_{\lambda,\mu} m_{\mu,\nu} = m_{\lambda,\nu[j]}$$

Using these formulae, we can replace the two summations over j and ν with one summation over $\gamma \Vdash k$ (so $\gamma = \nu[j]$). We obtain:

$$\dim L(\sigma, \lambda') = \sum_{\gamma \Vdash k} S_{(d^k)}(q) R_\gamma(q^d) m_{\lambda,\gamma}$$

which verifies the induction step. $\square$

5.5e. **Remarks.** (i) For the unipotent case $\sigma = 1$, it is interesting to "set $q = 1$" in the right hand side of the dimension formula in Theorem 5.5d. Observe that $S_\lambda(1)$ is 0 unless $\lambda = (1^n)$, when it is 1. So the right hand side of the expression in the theorem is equal simply to $m_{\lambda,(1^n)}$, the weight multiplicity of the (1^n)-weight space of the irreducible module $L_n(\lambda)$ for the classical Schur algebra $S_1(n,n)$. This is well-known to be same as the dimension $\dim D^{\lambda'}$ of the irreducible $F\Sigma_n$-module parametrized by the partition λ', or zero if λ is not p-restricted.

(ii) The idea in this section is extended in [Br] to give a similar result to Theorem 5.5d concerning the Brauer character values of the irreducible FG_n-modules at unipotent elements. In particular, it is shown there that

$$S_\lambda(q) = \sum_{\mu \vdash n} K^{-1}_{\lambda,\mu} \tilde{K}_{\mu',(1^n)}(q)$$

where $K^{-1} = (K^{-1}_{\lambda,\mu})$ is the inverse of the matrix of Kostka numbers of [M, I, (6.4)] and $\tilde{K} = (\tilde{K}_{\lambda,\mu}(q))$ is the matrix of renormalized Kostka-Foulkes polynomials as in [M, III, (7.11)].

(iii) For an application of Theorem 5.5d, see [BK$_2$].

Bibliography

[Ac] B. Ackermann, *Erweiterte Howlett-Lehrer Theorie*, Diploma thesis, Universität Stuttgart, 1999.

[AF] F. W. Anderson and K. R. Fuller, *Rings and categories of modules*, Springer-Verlag, 1974.

[A] M. Auslander, Representation theory of Artin algebras I, *Comm. Algebra* **1** (1974), 177–268.

[BLM] A. A. Beilinson, G. Lusztig, and R. MacPherson, A geometric setting for the quantum deformation of GL_n, *Duke Math. J.* **61** (1990), 655–677.

[B] M. Broué, Isométries de caractères et équivalences de Morita ou dérivées, *Publ. Math. IHES* **71** (1990), 45–63.

[BM] M. Broué and J. Michel, Blocs et series de Lusztig dans un groupe réductif fini, *J. reine angew. Math.* **395** (1989), 56–67.

[Br] J. Brundan, Unipotent Brauer character values of $GL_n(q)$ and the forgotten basis of the Hall algebra, preprint, University of Oregon, 1999.

[BK_1] J. Brundan and A. Kleshchev, Modular Littlewood-Richardson coefficients, *Math. Z.* **232** (1999), 287–320.

[BK_2] J. Brundan and A. Kleshchev, Lower bounds for degrees of irreducible Brauer characters of finite general linear groups, *J. Algebra*, to appear.

[C] R. W. Carter, *Finite groups of Lie type: conjugacy classes and complex characters*, Wiley, New York, 1985.

[Cl] G. Cliff, A tensor product theorem for quantum linear groups at even roots of unity, *J. Algebra* **165** (1994), 566–575.

[CPS_1] E. Cline, B. Parshall and L. Scott, Derived categories and Morita theory, *J. Algebra* **104** (1986), 397–409.

[CPS_2] E. Cline, B. Parshall and L. Scott, Finite dimensional algebras and highest weight categories, *J. reine angew. Math.* **391** (1988), 85–99.

[CPS_3] E. Cline, B. Parshall and L. Scott, Generic and q-rational representation theory, *Publ. RIMS Kyoto Univ.* **35** (1999), 31–90.

[DL] P. Deligne and G. Lusztig, Representations of reductive groups over finite fields, *Annals of Math.* **103** (1976), 103–161.

[DM_1] F. Digne and J. Michel, Foncteurs de Lusztig et caractères des groupes linéaires et unitaires sur un corps fini, *J. Algebra* **107** (1987), 217–255.

[DM_2] F. Digne and J. Michel, *Representations of finite groups of Lie type*, LMS Student Texts no. 21, Cambridge University Press, 1991.

[D_1] R. Dipper, On the decomposition numbers of finite general linear groups, *Trans. Amer. Math. Soc.* **290** (1985), 315–344.

[D_2] R. Dipper, On the decomposition numbers of finite general linear groups II, *Trans. Amer. Math. Soc.* **292** (1985), 123–133.

[D_3] R. Dipper, On quotients of Hom functors and representations of finite general linear groups I, *J. Algebra* **130** (1990), 235–259.

[D_4] R. Dipper, On quotients of Hom functors and representations of finite general linear groups II, *J. Algebra* **209** (1998), 199–269.

[DDo] R. Dipper and S. Donkin, Quantum GL_n, *Proc. London Math. Soc.* **63** (1991), 165–211.

[DDu_1] R. Dipper and J. Du, Harish-Chandra vertices, *J. reine angew. Math.* **437** (1993), 101–130.

[DDu_2] R. Dipper and J. Du, Harish-Chandra vertices and Steinberg's tensor product theorem for general linear groups in non-describing characteristic, *Proc. London Math. Soc.* **75** (1997), 559–599.

[DF] R. Dipper and P. Fleischmann, Modular Harish-Chandra theory I, *Math. Z.* **211** (1992), 49–71.

[DJ_1] R. Dipper and G. James, Representations of Hecke algebras of general linear groups, *Proc. London Math. Soc.* **52** (1986), 20–52.

[DJ_2] R. Dipper and G. James, Identification of the irreducible modular representations of $GL_n(q)$, *J. Algebra* **104** (1986), 266–288.

[DJ_3] R. Dipper and G. James, The q-Schur algebra, *Proc. London Math. Soc.* **59** (1989), 23–50.

[DJ_4] R. Dipper and G. James, q-Tensor space and q-Weyl modules, *Trans. Amer. Math. Soc.* **327** (1991), 251–282.

[DR] V. Dlab and C. M. Ringel, Quasi-hereditary algebras, *Illinois J. Math.* **33** (1989), 280–291.

[Do_1] S. Donkin, Hopf complements and injective comodules for algebraic groups, *Proc. London Math. Soc.* **40** (1980), 298–319.

[Do_2] S. Donkin, Finite resolutions of modules for reductive algebraic groups, *J. Algebra* **101** (1986), 473–488.

[Do_3] S. Donkin, On Schur algebras and related algebras I, *J. Algebra* **104** (1986), 310–328.

[Do_4] S. Donkin, On Schur algebras and related algebras II *J. Algebra* **111** (1987), 354–364.

[Do$_5$] S. Donkin, On tilting modules for algebraic groups, *Math. Z.* **212** (1993), 39–60.

[Do$_6$] S. Donkin, Standard homological properties of quantum GL_n, *J. Algebra* **181** (1996), 235–266.

[Do$_7$] S. Donkin, *The q-Schur algebra*, LMS Lecture Notes no. 253, Cambridge University Press, 1998.

[Du] J. Du, A note on quantized Weyl reciprocity at roots of unity, *Alg. Colloq.* **2** (1995), 363–372.

[FS] P. Fong and B. Srinavasan, The blocks of finite general linear and unitary groups, *Invent. Math.* **69** (1982), 109–153.

[GH] M. Geck and G. Hiss, Basic sets of Brauer characters of finite groups of Lie type, *J. reine angew. Math.* **418** (1991), 173–188.

[Ge] S. I. Gelfand, Representations of the full linear group over a finite field, *Math. USSR Sbornik* **12** (1970), 13–39.

[GG] I. M. Gelfand and M.I. Graev, The construction of irreducible representations of simple algebraic groups over a finite field, *Dokl. Akad. Nauk USSR* **147** (1962), 529–532.

[G$_1$] J. A. Green, The characters of the finite general linear groups, *Trans. Amer. Math. Soc.* **80** (1955), 402–447.

[G$_2$] J. A. Green, *Polynomial representations of* GL_n, Lecture Notes in Math., vol. 830, Springer-Verlag, 1980.

[Gr] R. M. Green, *q-Schur algebras and quantized enveloping algebras*, PhD thesis, Warwick University, 1995.

[GT] R. Guralnick and Pham Huu Tiep, Low-dimensional representations of special linear groups in cross characteristics, *Proc. London Math. Soc.* **78** (1999), 116–138.

[HL$_1$] R. Howlett and G. Lehrer, Induced cuspidal representations and generalized Hecke rings, *Invent. Math.* **58** (1980), 37–64.

[HL$_2$] R. Howlett and G. Lehrer, On Harish-Chandra induction for modules of Levi subgroups, *J. Algebra* **165** (1994), 172–183.

[HR] D. Happel and C. M. Ringel, Tilted algebras, *Trans. Am. Math. Soc.* **276** (1982), 399–443.

[J$_1$] G. James, *Representations of general linear groups*, LMS Lecture Notes no. 94, Cambridge University Press, 1984.

[J$_2$] G. James, The irreducible representations of the finite general linear groups, *Proc. London Math. Soc.* **52** (1986), 236–268.

[JM] G. James and A. Mathas, A q-analogue of the Jantzen-Schaper theorem, *Proc. London Math. Soc.* **74** (1997), 241–274.

[Ja] J. C. Jantzen, *Representations of Algebraic Groups*, Academic Press, Orlando, 1987.

[JS] J. C. Jantzen and G. M. Seitz, On the representation theory of the symmetric groups, *Proc. London Math. Soc.* **65** (1992), 475–504.

[Ka] G. Karpilovsky, *Group representations*, vol. 4, Math. Studies no. 182, North-Holland, Amsterdam, 1995.

[KK] M. Klucznik and S. König, *Characteristic tilting modules over quasi-hereditary algebras*, Lecture Notes, University of Bielefeld, 1999.

[La] P. Landrock, *Finite Group Algebras and their Modules*, Cambridge University Press, 1983.

[L1] G. Lusztig, On the finiteness of the number of unipotent classes, *Invent. Math.* **34** (1976), 201–213.

[L2] G. Lusztig, Finite dimensional Hopf algebras arising from quantized universal enveloping algebras, *J. Amer. Math. Soc.* **3** (1990), 257–297.

[LS] G. Lusztig and B. Srinivasan, The characters of the finite unitary groups, *J. Algebra* **49** (1977), 167–171.

[M] I. G. Macdonald, *Symmetric functions and Hall polynomials*, Oxford Mathematical Monographs, second edition, OUP, 1995.

[Ma] A. Mathas, *Hecke algebras and Schur algebras of the symmetric group*, Lecture notes, Universität Bielefeld, 1998.

[MP] O. Mathieu and G. Papadopoulo, A character formula for a family of simple modular representations of GL_n, *Comment. Math. Helv.* **74** (1999), 280–296.

[PW] B. Parshall and J.-P. Wang, Quantum linear groups, *Mem. Amer. Math. Soc.* **439** (1991).

[R] C. M. Ringel, The category of modules with good filtrations over a quasi-hereditary algebra has almost split sequences, *Math. Z.* **208** (1991), 209–225.

[S] V. Schubert, *Durch den Hom-Funktor induzierte Äquivalenzen*, PhD thesis, Universität Stuttgart, 1999.

[Se] J.-P. Serre, *Linear representations of finite groups*, Gradute Texts in Mathematics 42, Springer-Verlag, Berlin, 1977.

[Sw] M. Sweedler, *Hopf algebras*, Benjamin, New York, 1969.

[T] M. Takeuchi, The group ring of $GL_n(q)$ and the q-Schur algebra, *J. Math. Soc. Japan* **48** (1996), 259–274.

[Th] E. Thoma, Die Einschränkung der Charactere von $GL(n, q)$ auf $GL(n-1, q)$, *Math. Z.* **119** (1971), 321–338.

[Z] A. Zelevinsky, *Representations of finite classical groups*, Lecture Notes in Math. 869, Springer-Verlag, Berlin, 1981.

Editorial Information

To be published in the *Memoirs*, a paper must be correct, new, nontrivial, and significant. Further, it must be well written and of interest to a substantial number of mathematicians. Piecemeal results, such as an inconclusive step toward an unproved major theorem or a minor variation on a known result, are in general not acceptable for publication. Papers appearing in *Memoirs* are generally longer than those appearing in *Transactions*, which shares the same editorial committee.

As of September 30, 2000, the backlog for this journal was approximately 11 volumes. This estimate is the result of dividing the number of manuscripts for this journal in the Providence office that have not yet gone to the printer on the above date by the average number of monographs per volume over the previous twelve months, reduced by the number of volumes published in four months (the time necessary for preparing a volume for the printer). (There are 6 volumes per year, each containing at least 4 numbers.)

A Consent to Publish and Copyright Agreement is required before a paper will be published in the *Memoirs*. After a paper is accepted for publication, the Providence office will send a Consent to Publish and Copyright Agreement to all authors of the paper. By submitting a paper to the *Memoirs*, authors certify that the results have not been submitted to nor are they under consideration for publication by another journal, conference proceedings, or similar publication.

Information for Authors

Memoirs are printed from camera copy fully prepared by the author. This means that the finished book will look exactly like the copy submitted.

The paper must contain a *descriptive title* and an *abstract* that summarizes the article in language suitable for workers in the general field (algebra, analysis, etc.). The *descriptive title* should be short, but informative; useless or vague phrases such as "some remarks about" or "concerning" should be avoided. The *abstract* should be at least one complete sentence, and at most 300 words. Included with the footnotes to the paper should be the 2000 *Mathematics Subject Classification* representing the primary and secondary subjects of the article. The classifications are accessible from `www.ams.org/msc/`. The list of classifications is also available in print starting with the 1999 annual index of *Mathematical Reviews*. The Mathematics Subject Classification footnote may be followed by a list of *key words and phrases* describing the subject matter of the article and taken from it. Journal abbreviations used in bibliographies are listed in the latest *Mathematical Reviews* annual index. The series abbreviations are also accessible from `www.ams.org/publications/`. To help in preparing and verifying references, the AMS offers MR Lookup, a Reference Tool for Linking, at `www.ams.org/mrlookup/`. When the manuscript is submitted, authors should supply the editor with electronic addresses if available. These will be printed after the postal address at the end of the article.

Electronically prepared manuscripts. The AMS encourages electronically prepared manuscripts, with a strong preference for $\mathcal{A}_{\mathcal{M}}\mathcal{S}$-LaTeX. To this end, the Society has prepared $\mathcal{A}_{\mathcal{M}}\mathcal{S}$-LaTeX author packages for each AMS publication. Author packages include instructions for preparing electronic manuscripts, the *AMS Author Handbook*, samples, and a style file that generates the particular design specifications of that publication series. Though $\mathcal{A}_{\mathcal{M}}\mathcal{S}$-LaTeX is the highly preferred format of TeX, author packages are also available in $\mathcal{A}_{\mathcal{M}}\mathcal{S}$-TeX.

Authors may retrieve an author package from e-MATH starting from `www.ams.org/tex/` or via FTP to `ftp.ams.org` (login as `anonymous`, enter username as password, and type `cd pub/author-info`). The *AMS Author Handbook* and the *Instruction Manual* are available in PDF format following the author packages link from `www.ams.org/tex/`. The author package can be obtained free of charge by sending email to `pub@ams.org` (Internet) or from the Publication Division, American Mathematical Society, P.O. Box 6248, Providence, RI 02940-6248. When requesting an author package, please specify $\mathcal{A}_{\mathcal{M}}\mathcal{S}$-LaTeX or $\mathcal{A}_{\mathcal{M}}\mathcal{S}$-TeX, Macintosh or IBM (3.5) format, and the publication in which your paper will appear. Please be sure to include your complete mailing address.

Sending electronic files. After acceptance, the source file(s) should be sent to the Providence office (this includes any TeX source file, any graphics files, and the DVI or PostScript file).

Before sending the source file, be sure you have proofread your paper carefully. The files you send must be the EXACT files used to generate the proof copy that was accepted for publication. For all publications, authors are required to send a printed copy of their paper, which exactly matches the copy approved for publication, along with any graphics that will appear in the paper.

TeX files may be submitted by email, FTP, or on diskette. The DVI file(s) and PostScript files should be submitted only by FTP or on diskette unless they are encoded properly to submit through email. (DVI files are binary and PostScript files tend to be very large.)

Electronically prepared manuscripts can be sent via email to `pub-submit@ams.org` (Internet). The subject line of the message should include the publication code to identify it as a Memoir. TeX source files, DVI files, and PostScript files can be transferred over the Internet by FTP to the Internet node `e-math.ams.org` (130.44.1.100).

Electronic graphics. Comprehensive instructions on preparing graphics are available at `www.ams.org/jourhtml/graphics.html`. A few of the major requirements are given here.

Submit files for graphics as EPS (Encapsulated PostScript) files. This includes graphics originated via a graphics application as well as scanned photographs or other computer-generated images. If this is not possible, TIFF files are acceptable as long as they can be opened in Adobe Photoshop or Illustrator. No matter what method was used to produce the graphic, it is necessary to provide a paper copy to the AMS.

Authors using graphics packages for the creation of electronic art should also avoid the use of any lines thinner than 0.5 points in width. Many graphics packages allow the user to specify a "hairline" for a very thin line. Hairlines often look acceptable when proofed on a typical laser printer. However, when produced on a high-resolution laser imagesetter, hairlines become nearly invisible and will be lost entirely in the final printing process.

Screens should be set to values between 15% and 85%. Screens which fall outside of this range are too light or too dark to print correctly. Variations of screens within a graphic should be no less than 10%.

Inquiries. Any inquiries concerning a paper that has been accepted for publication should be sent directly to the Electronic Prepress Department, American Mathematical Society, P. O. Box 6248, Providence, RI 02940-6248.

Editors

This journal is designed particularly for long research papers (and groups of cognate papers) in pure and applied mathematics. Papers intended for publication in the *Memoirs* should be addressed to one of the following editors. In principle the Memoirs welcomes electronic submissions, and some of the editors, those whose names appear below with an asterisk (*), have indicated that they prefer them. However, editors reserve the right to request hard copies after papers have been submitted electronically. Authors are advised to make preliminary email inquiries to editors about whether they are likely to be able to handle submissions in a particular electronic form.

Algebra to CHARLES CURTIS, Department of Mathematics, University of Oregon, Eugene, OR 97403-1222 email: `cwc@darkwing.uoregon.edu`

Algebraic geometry and commutative algebra to LAWRENCE EIN, Department of Mathematics, University of Illinois, 851 S. Morgan (M/C 249), Chicago, IL 60607-7045; email: `ein@uic.edu`

Algebraic topology and cohomology of groups to STEWART PRIDDY, Department of Mathematics, Northwestern University, 2033 Sheridan Road, Evanston, IL 60208-2730; email: `priddy@math.nwu.edu`

Combinatorics and Lie theory to PHILIP J. HANLON, Department of Mathematics, University of Michigan, Ann Arbor, Michigan 48109-1003; email: `hanlon@math.lsa.umich.edu`

Complex analysis and complex geometry to DANIEL M. BURNS, Department of Mathematics, University of Michigan, Ann Arbor, MI 48109-1003; email: `dburns@math.lsa.umich.edu`

***Differential geometry and global analysis** to CHUU-LIAN TERNG, Department of Mathematics, Northeastern University, Huntington Avenue, Boston, MA 02115-5096; email: `terng@neu.edu`

***Dynamical systems and ergodic theory** to ROBERT F. WILLIAMS, Department of Mathematics, University of Texas, Austin, Texas 78712-1082; email: `bob@math.utexas.edu`

Geometric topology, knot theory, hyperbolic geometry, and general topoogy to JOHN LUECKE, Department of Mathematics, University of Texas, Austin, TX 78712-1082; email: `luecke@math.utexas.edu`

Harmonic analysis, representation theory, and Lie theory to ROBERT J. STANTON, Department of Mathematics, The Ohio State University, 231 West 18th Avenue, Columbus, OH 43210-1174; email: `stanton@math.ohio-state.edu`

***Logic** to THEODORE SLAMAN, Department of Mathematics, University of California, Berkeley, CA 94720-3840; email: `slaman@math.berkeley.edu`

Number theory to MICHAEL J. LARSEN, Department of Mathematics, Indiana University, Bloomington, IN 47405; email: `larsen@math.indiana.edu`

Operator algebras and functional analysis to BRUCE E. BLACKADAR, Department of Mathematics, University of Nevada, Reno, NV 89557; email: `bruceb@math.unr.edu`

***Ordinary differential equations, partial differential equations, and applied mathematics** to PETER W. BATES, Department of Mathematics, Brigham Young University, 292 TMCB, Provo, UT 84602-1001; email: `peter@math.byu.edu`

***Partial differential equations and applied mathematics** to BARBARA LEE KEYFITZ, Department of Mathematics, University of Houston, 4800 Calhoun Road, Houston, TX 77204-3476; email: `keyfitz@uh.edu`

***Probability and statistics** to KRZYSZTOF BURDZY, Department of Mathematics, University of Washington, Box 354350, Seattle, Washington 98195-4350; email: `burdzy@math.washington.edu`

***Real and harmonic analysis and geometric partial differential equations** to WILLIAM BECKNER, Department of Mathematics, University of Texas, Austin, TX 78712-1082; email: `beckner@math.utexas.edu`

All other communications to the editors should be addressed to the Managing Editor, WILLIAM BECKNER, Department of Mathematics, University of Texas, Austin, TX 78712-1082; email: `beckner@math.utexas.edu`.

Selected Titles in This Series

(Continued from the front of this publication)

677 **Volodymyr V. Lyubashenko,** Squared Hopf algebras, 1999

676 **S. Strelitz,** Asymptotics for solutions of linear differential equations having turning points with applications, 1999

675 **Michael B. Marcus and Jay Rosen,** Renormalized self-intersection local times and Wick power chaos processes, 1999

674 **R. Lawther and D. M. Testerman,** A_1 subgroups of exceptional algebraic groups, 1999

673 **John Lott,** Diffeomorphisms and noncommutative analytic torsion, 1999

672 **Yael Karshon,** Periodic Hamiltonian flows on four dimensional manifolds, 1999

671 **Andrzej Rosłanowski and Saharon Shelah,** Norms on possibilities I: Forcing with trees and creatures, 1999

670 **Steve Jackson,** A computation of δ_5^1, 1999

669 **Seán Keel and James McKernan,** Rational curves on quasi-projective surfaces, 1999

668 **E. N. Dancer and P. Poláčik,** Realization of vector fields and dynamics of spatially homogeneous parabolic equations, 1999

667 **Ethan Akin,** Simplicial dynamical systems, 1999

666 **Mark Hovey and Neil P. Strickland,** Morava K-theories and localisation, 1999

665 **George Lawrence Ashline,** The defect relation of meromorphic maps on parabolic manifolds, 1999

664 **Xia Chen,** Limit theorems for functionals of ergodic Markov chains with general state space, 1999

663 **Ola Bratteli and Palle E. T. Jorgensen,** Iterated function systems and permutation representation of the Cuntz algebra, 1999

662 **B. H. Bowditch,** Treelike structures arising from continua and convergence groups, 1999

661 **J. P. C. Greenlees,** Rational S^1-equivariant stable homotopy theory, 1999

660 **Dale E. Alspach,** Tensor products and independent sums of $\mathcal{L}_p$-spaces, $1 < p < \infty$, 1999

659 **R. D. Nussbaum and S. M. Verduyn Lunel,** Generalizations of the Perron-Frobenius theorem for nonlinear maps, 1999

658 **Hasna Riahi,** Study of the critical points at infinity arising from the failure of the Palais-Smale condition for n-body type problems, 1999

657 **Richard F. Bass and Krzysztof Burdzy,** Cutting Brownian paths, 1999

656 **W. G. Bade, H. G. Dales, and Z. A. Lykova,** Algebraic and strong splittings of extensions of Banach algebras, 1999

655 **Yuval Z. Flicker,** Matching of orbital integrals on $GL(4)$ and $GSp(2)$, 1999

654 **Wancheng Sheng and Tong Zhang,** The Riemann problem for the transportation equations in gas dynamics, 1999

653 **L. C. Evans and W. Gangbo,** Differential equations methods for the Monge-Kantorovich mass transfer problem, 1999

652 **Arne Meurman and Mirko Primc,** Annihilating fields of standard modules of $\mathfrak{sl}(2,\mathbb{C})^{\sim}$ and combinatorial identities, 1999

651 **Lindsay N. Childs, Cornelius Greither, David J. Moss, Jim Sauerberg, and Karl Zimmermann,** Hopf algebras, polynomial formal groups, and Raynaud orders, 1998

650 **Ian M. Musson and Michel Van den Bergh,** Invariants under Tori of rings of differential operators and related topics, 1998

649 **Bernd Stellmacher and Franz Georg Timmesfeld,** Rank 3 amalgams, 1998

648 **Raúl E. Curto and Lawrence A. Fialkow,** Flat extensions of positive moment matrices: Recursively generated relations, 1998

For a complete list of titles in this series, visit the AMS Bookstore at **www.ams.org/bookstore/**.